SEVENTH EDITION

EXERCISES IN HELPING SKILLS
A MANUAL TO ACCOMPANY
THE SKILLED HELPER

Gerard Egan

Professor Emeritus
Loyola University of Chicago

with assistance from
Rich McGourty

BROOKS/COLE

—✳—
™
THOMSON LEARNING

Australia • Canada • Mexico • Singapore • Spain • United Kingdom • United States

BROOKS/COLE

THOMSON LEARNING ™

Assistant Editor: *Alma Dea Michelena*
Editorial Assistant: *Sheila Walsh*
Production Coordinator: *Dorothy Bell*
Cover Design: *Andrew Ogus*

Cover Illustration: *Ryuichi Okano/Photonica*
Print Buyer: *Micky Lawler*
Printing and Binding: *Webcom Limited*

For more information about this or any other Brooks/Cole product, contact:
BROOKS/COLE
511 Forest Lodge Road
Pacific Grove, CA 93950 USA
www.brookscole.com
1-800-423-0563 (Thomson Learning Academic Resource Center)

Printed in Canada

10 9 8 7 6 5 4 3 2

ISBN 0-534-36732-1

CONTENTS

The exercises in this manual accompany the seventh edition of *The Skilled Helper* by Gerard Egan (2002). The six parts of the manual correspond to the six major parts of the text. The chapters of the manual correspond to the chapters of the book. The exercises themselves are numbered consecutively throughout the text.

INTRODUCTION

Getting the Most Out of These Exercises

These exercises are meant to help you translate the concepts in *The Skilled Helper* into skills that you can use. You may already possess some of the skills. If so, those skills can be improved. Skills that are less developed in you will require persistence.

There is no real substitute for practice and repetition. This is why learning with the help of others in your class is so important. With their help you will have more opportunities to practice and receive feedback. To assist you in this effort the exercises have been designed to provide a wide range of examples. You will also find a blend of in-class exercises and other exercises which take place in the settings of your life.

A TRAINING PROGRAM FOR HELPERS

As the title *Skilled Helper* suggests, helpers-to-be need, not just a helping model, but a range of techniques and skills to make the model work. They include basic and advanced communication skills, the ability to establish working relationships with clients, and the skills needed to help clients identify and explore problem situations, identify and explore unused opportunities, set problem-managing and opportunity-developing goals, develop action plans to accomplish these goals, implement plans, and engage in ongoing evaluation of this entire process. These skills are not just bits of helping technology but rather creative and humane ways of helpers' being with clients in the their efforts to better their lives. The only way to acquire these skills is by learning them experientially, practicing them, and using them until they become second nature.

You are about to embark on a training program in counseling. The following are standard steps you will encounter in such a program:

1. **Cognitive Understanding**. Develop a cognitive understanding of a particular helping method or the skill of delivering it. You can do this by reading the text and listening to lectures.
2. **Clarification**. Clarify what you have read or heard. This can be done through instructor-led questioning and discussion.

The desired outcome of Steps 1 and 2 is cognitive clarity.

3. **Modeling**. Watch experienced instructors model the skill or method in question. This can be done "live" or through films and videotapes.
4. **Written Exercises**. Do the exercises in this manual that are related to the skill or method you are learning. The purpose of this is to demonstrate to yourself that you understand the helping method or skill enough to begin to practice it. The exercises in this manual are a way of practicing the skills and methods "in private" before practicing them with your fellow trainees.

The desired outcome of Steps 3 and 4 is behavioral clarity.

5. **Practice**. Move into smaller groups to practice the skill or method in question with your fellow trainees.

6. **Feedback**. During these practice sessions evaluate your own performance and get feedback from a trainer and from your fellow trainees. This feedback serves to confirm what you are doing right and to correct what you are doing wrong. The use of videotape to provide feedback is very helpful.

**The desired outcome of Steps 5 and 6 is initial competence
in using both the model and the skills that make it work.**

7. **Evaluating the Learning Experience**. From time to time stop and reflect on the training process itself. Take the opportunity to express how you feel about the learning program and how you feel about your own progress. While Steps 1 through 6 deal with the task of learning the helping model and the methods and skills that make it work, Step 7 deals with developing and maintaining a learning community.
8. **Supervised Practice with Actual Clients**. Finally, when it is deemed that you are ready, apply what you have learned to actual clients under supervision. Supervision is an extremely important part of the learning process.

The desired outcome of Steps 7 and 8 is initial mastery.

The program in which you are enrolled may cover only a few of these steps. Comprehensive programs for training professional helpers, however, should eventually include all steps, even though not in the order presented here. Of course, the best helpers continue to learn throughout their careers--from their own experiences, from the experiences of their colleagues, from reading, from seminars, from formal and informal research, and, perhaps most importantly, from their interactions with their clients. Full mastery, therefore, is a journey rather than a destination.

Extensive self-exploration exercises are included in this manual to help you take a good look at your own strengths and weaknesses as a helper. As such they provide a way of having you apply the helping model to yourself first before trying it out on others. You can use these exercises to confirm strengths that enable you to be with clients effectively and manage weaknesses that would stand in the way of helping clients manage problem situations.

SHARING YOUR OWN EXPERIENCE

As you will see the exercises provide many opportunities to share your own experience. If you can do this in a way that is a good fit for the pace of learning in your class and yourself, you will find that the learning group you are in will become a safe and productive place for learning.

- **Achieve a Balance Between Comfort and Risk-taking.** At no time should you feel you are talking about yourself in a way that makes you uncomfortable. No one can learn if they are ill at ease or if they feel coerced. At the same time the quality of learning improves if each learner contributes something of themselves that is real and substantial.

- **Dealing With Real Issues**. Role-playing, while not easy, is still less personally demanding than discussing your own real concerns in practice sessions. Under certain conditions, the training process can be used to look at some of the real problems or concerns in your own life, *especially issues that relate to your effectiveness as a helper*. For instance, if you tend to be an impatient person--one who places unreasonable demands on others--you will have to examine and change this behavior if you want to become an effective helper. Or, if you are very nonassertive, this may inhibit you in helping clients challenge themselves.

- **Use Real Material.** One way to make sure that the exchanges in class are compelling is to draw upon your real experience. Another reason for using real problems or concerns when you take the role of the client is that it gives you some experience of **being** a client. Then, when you face real clients, you can appreciate some of the misgivings they might have in talking to a relative stranger about the intimate details of their lives. Other things being equal, I would personally prefer going to a helper who has had some kind of personal experience as a client. There are several exercises in the book to help you identify what you can talk about.

- **Use Role Play as Needed.** An alternative to using your real experience is the use of role play. In experiential programs you are going to be asked to act both as helper and as client in practice sessions. In the written exercises in this manual, you are asked at one time or another to play each of these roles. Role playing is not easy. It forces you to get "inside" the clients you are playing and understand their concerns as they experience them. This can help you become more empathic. An over-reliance on role play is not conducive to learning. It can make skills training look more like a theater class. That said, there may be times when skills can be practiced under a wider variety of conditions if the members of the class use some role play.

- **Observe Confidentiality.** A reminder about confidentiality might seem unnecessary. It is so important, however, that it cannot be repeated too often. The information that is shared in the training program is not to be discussed outside of the class and even then you may have boundaries distinguishing what is said in small group and what can be discussed as a whole class. Think of this as practice for when you are a counselor and your clients entrust you with their stories.

GIVING FEEDBACK TO SELF AND OTHERS

If you are working with a learning partner or if you are in an experiential training group, you will be called on to give feedback both to yourself and to your co-learners on how well you are learning and using helping methods and skills. Giving feedback well is an art. Here are some guidelines to help you develop that art.

1. **Keep the goal of feedback in mind.** In giving feedback, always keep the generic goal of feedback in mind which is to help the other person (or yourself, in the case of self-feedback) do a better job. Improved performance is the desired outcome. Applied to helping, this means providing the kind of feedback to yourself and your co-learners that will help you become better helpers. Feedback will help you learn every stage and step of the model.

2. **Give positive feedback.** Tell your co-learners what they are doing well. This reinforces useful helping behaviors. "You leaned toward him and kept good eye contact even though he became very intense and emotional. Your behavior sent the right I'm-still-with-you message."

3. **Don't avoid corrective feedback.** To learn from our mistakes we must know what they are. Corrective feedback given in a caring way is a powerful tool for learning. "You seem reluctant to challenge your clients. For instance, Sam [the client] didn't fulfill his contract from the last meeting and you let it go. You fidgeted when he said he didn't get around to it."

4. **Be specific.** General statements like, "I liked your style in challenging your client," or "You could have been more understanding," are not helpful. Change them to specific descriptions such as, "Your challenge was helpful because you pointed out how self-defeating her internal conversations with herself are and you hinted at ways she could change these conversations."

5. **Focus on behavior rather than traits.** Point out what the helper does or fails to do. Do not focus on traits or use labels such as, "You showed yourself to be a leader." or "You're still a bumbler." Avoid using negative traits such as "lazy," "a slow learner," "incompetent," "manipulative," and so forth. This is just name-calling and creates a negative learning climate in the group. For example, the following

statement deals with specific behaviors rather than traits. "You didn't become defensive when he criticized you; he listened to you better after that."

6. **Indicate the impact of the behavior on the client.** Feedback should help counselors-to-be interact more productively with clients. It helps, then, to indicate the impact of the helper's behavior on the client. "You interrupted the client three times in the space of about two minutes. After the third time, she spoke less intensely and switched to safer topics. She seemed to wander around."

7. **Provide hints for better performance.** Often helpers, once they receive corrective feedback, know how to change their behavior. After all, they are learning how to help in the training program. Sometimes, however, if your fellow trainee agrees with the feedback but does not know how to change his or her behavior, suggestions or hints on how to improve performance are useful. These, too, should be specific and clear. "You are having trouble providing your clients with empathic highlights because you allow them to talk too long. When they go on and on, they make so many points that you don't know which to respond to. Try interrupting your clients gently so that you can respond to key messages as they come up."

8. **Be brief.** Feedback that is both specific and brief is most helpful. Long-winded feedback proves to be a waste of time. A helper might need feedback on a number of points. In this case, provide feedback on one or two points. Give further feedback later.

9. **Use dialogue.** Feedback is more effective if it takes place through a dialogue between the giver and receiver--a brief dialogue, of course. This gives the receiver an opportunity to clarify what the feedback giver means and to ask for suggestions if he or she needs them. A dialogue helps the receiver better "own" the feedback.

TRANSFER LEARNING FROM THE CLASSROOM TO THE REAL WORLD.

The true acquisition of the helping skills is accomplished when they are at your command in everyday living and not just when you are intentionally practicing them in a training program. That is why many exercises are designed to encourage you to take the skills you learn in the classroom into regular day-to-day living. It is a missed opportunity if your practice is confined to the classroom.

PART ONE

LAYING THE GROUNDWORK

Although the centerpiece of this book is a problem-management and

opportunity-development helping model and the methods and

communication skills that make it work, there is some groundwork to be

laid. This includes outlining the nature and goals of helping (Chapter 1),

providing an overview of the helping process itself (Chapter 2), and

describing the helping relationship and the values that should drive it

(Chapter3).

Chapter 1

INTRODUCTION

You may or may not intend to become a professional helper. Whether you do or not, learning the model, methods, and skills of *The Skilled Helper* can help you become more effective in your interactions with both yourself and others in all the social settings of life including family, friendship groups, and work settings. It is important that you understand what helping is all about in order to get the most out of doing these exercises in this manual
.

EXERCISE 1: UNDERSTANDING WHAT HELPING IS ALL ABOUT

1. Read the sections in *The Skilled Helper* dealing with what helping is all about, the goals of helping, and helping as a learning process.
2. Picture yourself as a helper talking to a prospective client.
3. On a separate piece of paper write a brief description of what helping is and what its goals are. Use your own words. As you write, see yourself talking with the client.
4. Share what you have written with a learning partner. Note the similarities and the differences between the two statements. To what degree do they include the main ideas outlined in the text? In what way do you think you should modify your statement?

EXERCISE 2: PREPARING TO PLAY THE ROLE OF CLIENT: STRENGTHS AND SOFT SPOTS IN MY LIFE

Here, in quotation form, are some of the kinds of problems, issues, and concerns that trainees have dealt with during training programs. As you read them, check (✓) the ones that pertain to yourself.

- I'm shy. My shyness takes the form of being afraid to meet strangers and being afraid to reveal myself to others.
- I'm a fairly compliant person. Others can push me around and get away with it.
- I get angry fairly easily and let my anger spill out on others in irresponsible ways. I think my anger is often linked to not getting my own way.
- I'm a somewhat lazy person. I find it especially difficult to expend the kind of energy necessary to listen to and get involved with others.
- I'm somewhat fearful of persons of the opposite sex. This is especially true if I think they are putting some kind of demand on me for closeness. I get nervous and put them off.
- I'm a rather insensitive person, or so I have been told. I'm a kind of bull-in-the-china-shop type. Not much tact.
- I'm overly controlled. I don't let my emotions show very much. Sometimes I don't even want to know what I'm feeling myself.
- I like to control others, but I like to do so in subtle ways. I want to stay in charge of interpersonal relationships at all times.
- I have a strong need to be liked by others. I seldom do anything that might offend others or that others would not approve of. I have a very strong need to be accepted.
- I have few positive feelings about myself. I put myself down in a variety of ways. I get depressed a lot.
- I never stop to examine my values. I think I hold some conflicting values. I'm not even sure why I'm interested in becoming a helper.
- I feel almost compelled to help others. It's part of my religious background. It's as if I didn't even have a choice.
- I'm sensitive and easily hurt. I think I send out messages to others that say, "be careful of me."
- I'm overly dependent on others. My self-image depends too much on what others think of me.
- A number of people see me as a "difficult" person. I'm highly individualistic. I'm ready to fight if anyone imposes on my freedom.
- I'm anxious a lot of the time. I'm not even sure why. My palms sweat a lot in interpersonal situations.
- I see myself as a rather colorless, uninteresting person. I'm bored with myself at times and I assume that others are bored with me.
- I'm somewhat irresponsible. I take too many risks, especially risks that involve others. I'm very impulsive. That's probably a nice way of saying that I lack self-control.
- I'm very stubborn. I have fairly strong opinions. I argue a lot and try to get others to see things my way. I argue about very little things.
- I don't examine myself or my behavior very much. I'm usually content with the way things are. I don't expect too much of myself or of others.
- I can be sneaky in my relationships with others. I seduce people in different ways—not necessarily sexual—by my "charm." I get them to do what I want.
- I like the good life. I'm pretty materialistic and I like my own comfort. I don't often go out of my way to meet the needs of others.
- I'm somewhat lonely. I don't think others like me, if they think about me at all. I spend time feeling sorry for myself.

- I'm awkward in social situations. I don't do the right thing at the right time. I don't know what others are feeling when I'm with them and I guess I seem callous.
- Others see me as "out of it" a great deal of the time. I guess I am fairly naive. Others seem to have deeper or more interesting experiences than I do. I think I've grown up too sheltered.
- I'm stingy with both money and time. I don't want to share what I have with others. I'm pretty selfish.
- I'm somewhat of a coward. I sometimes find it hard to stand up for my convictions even when I meet light opposition. It's easy to get me to retreat from my opinions.
- I hate conflict. I'm more or less a peace-at-any-price person. I run when things get heated up.
- I don't like it when others tell me I'm doing something wrong. I usually feel attacked and I fall silent or attack back.

Using this list to "prime the pump," jot down three issues or concerns that you have. Choose items that might stand in the way of being an effective helper. Try to name issues that have some substance but which you would be willing to discuss in the training group.

Issue 1.

_____

_____

_____

Issue 2.

_____

_____

_____

Issue 3.

_____

_____

_____

EXERCISE 3: TURNING PROBLEMS INTO OPPORTUNITIES

While on the surface the long list of issues or concerns looks like a list of problems, it can also be seen as a list of opportunities. Take the last item, one that Helen had on her list.

> "I don't like it when others tell me I'm doing something wrong. I usually feel attacked and I fall silent or attack back."

Once Helen recognizes that she reacts poorly to negative or corrective feedback, she has the opportunity of changing her behavior. If she works at it, there will come a time when she will be able to say:

> "When people give me negative or corrective feedback, I use it as an opportunity to learn something about myself. Even when what the other person is saying is unfair, I can listen and learn. And it's an opportunity for self-discipline. I ask questions to get at the heart of

what the person is saying. It's not always easy and I don't always succeed, but I keep learning about myself."

You can do the same with the three issues you have chosen. Take each issue and, like Helen, turn each problem into an opportunity.

Opportunity 1.

_____

_____

_____

Opportunity 2.

_____

_____

_____

Opportunity 3.

_____

_____

_____

EXERCISE 4: STRENGTHS AND SOFT SPOTS

One way of preparing yourself to talk about issues that can influence the quality of your helping is to review both strengths and soft spots in your life. All of us have strengths, things we do well, and soft spots, areas in which we could use some improvement. I might well be a decisive person and that is a strength for me as a helper but, in being decisive, I might push others too hard and that is a soft spot needing improvement. In this case, a strength pushed too hard becomes a soft spot.

1.   In this exercise, merely jot down on separate paper—in whatever way they come to you—attitudes, skills, and behaviors that you have that help you lead a fuller life (strengths). Also list things that need some improvement (soft spots). In order to stress the positive, see if you can write down at least two things that are going right for each thing that needs improvement. Read the list in the example that follows and then do your own. Don't worry whether the problems or concerns you list are really important. Jot down whatever comes to your mind. Your own list may include items similar to those in the example, but it may be quite different because it will reflect you and not someone else.

Example

	Strengths		Soft Spots

Strengths

- I have lots of friends.
- Others can count on me.
- I've got a good part-time job; they like the quality of my work.
- I'm fairly smart.
- I am very healthy.
- My belief in God gives me a center.
- I'm hard working.
- I have a mind of my own; I don't give in just to please others

Soft Spots

- I am extremely demanding of myself.
- I get dependent on others too easily.
- I am slow to take risks even when they are reasonable.
- I can be arrogant with people who are not as smart as I am.

2. Review your list once you are finished. Star one or two items in each column that are the most important for you.
3. Share your list with a learning partner and discuss how you would like to make better use of the starred strengths and what you would like to do about the starred soft spots.
4. Keep this list. It can help you choose issues to discuss when you play the role of client.

EXERCISE 5: SPOTTING UNUSED OPPORTUNITIES

Unused opportunities are as important as problems. This exercise helps you take a look at some unused opportunities in your own life which, if exploited, well might help you become a better counselor.

1. Return to the strengths part of Exercise 4.
2. Read your list and identify two ways you might leverage them into opportunities that would contribute to your being a more effective helper.
3. Share the opportunities you find and talk them through with a learning partner.

Example. Sonya is a part-time student in a counseling program. She also has a job that consumes 30 hours of her time per week. She reviews her list of strengths and says, "I'm very hard working both at school and at work. I'm also very careful in how I do my work. However, I don't put the same effort into developing my social life. I could apply these two strengths in developing a fuller social life. After all, helping is about people and I need to get better at people skills." She goes on to search for a second unused opportunity. In talking them through with a learning partner, she comes to realize that she can use the training group as a "lab" for developing a better social life.

Chapter 2

AN OVERVIEW OF THE HELPING PROCESS

In Chapter One you were asked to explain the overall goals of counseling to another person. Chapter Two goes into more detail about the helping process itself. Do this exercise **before** reading Chapter Two or the outline of the helping process below.

EXERCISE 6: EXPLAINING WHAT HELPING INVOLVES.

1.   Before reading Chapter Two or the summary below, explain to a friend (not your learning partner) what helping involves. Assume this person is considering going to a counselor and knows little about what to expect other than, "You get in there and you talk about yourself."
2.   After reading Chapter Two, chose a different friend and this time explain what helping involves in light of your new knowledge. Use your own words rather than the jargon of the text.
3.   Discuss both experiences with your learning partner. Consider the following questions:

- What did you emphasize in each of the two efforts?
- What aspects of helping were omitted in your first effort?
- How did you deal with opportunity development?
- How did you describe the helper's role? The client's role?
- How did your friends' reactions to the two efforts differ?

THE STAGES AND STEPS OF THE HELPING PROCESS

Here is a brief outline of the skilled-helper model. There is, of course, a fuller outline in Chapter Two of *The Skilled Helper*.

STAGE I: "WHAT'S GOING ON?"
CLARIFICATION OF THE KEY ISSUES
CALLING FOR CHANGE

Clients can neither manage problem situations nor develop unused opportunities unless they identify and understand them. Exploration and clarification of problems and opportunities take place in Stage I. This stage deals with the current state of affairs, that is, the problem situations or unused opportunities that prompt clients to come for help in the first place. This stage includes the following steps:

I-A. Help clients tell their stories. First of all, clients need to talk about their problems and concerns, that is, tell their stories. Some do so easily, others with a great deal of difficulty. You need to develop a set of attitudes and communication skills that will enable you to help clients reveal problems in living and unused potential. This means helping clients find out what's going wrong and what's going right in their lives.

I-B. Help clients challenge their blind spots and develop new perspectives. This means helping clients see themselves, their concerns, and the contexts of their concerns more objectively. This enables clients to understand not only their problems and unused opportunities, but also ways in which they want their lives to be different. Your ability to help clients challenge themselves at every stage and step of the helping process can add a great deal of value.

I-C. Help clients focus on issues that will make a difference. This means helping clients identify their most important concerns, especially if they have a number of problems. Effective counselors help clients work on "high-leverage" issues, that is, problems and opportunities that will make a difference in clients' lives.

STAGE II: "WHAT DOES SUCCESS LOOK LIKE?"
HELPING CLIENTS DETERMINE OUTCOMES:
GOALS, ACCOMPLISHMENTS, IMPACT

Once clients understand either problem situations or unused opportunities more clearly, they often need help in determining what they want instead of what they have. They need to develop a picture of a better future, choose specific goals to work on, and commit themselves to a program of constructive change. For instance, at Stage II you might be helping a troubled married couple outline what a better marital relationship might look like.

II-A. "What do you want?" Help clients develop a range of possibilities for a better future. If a client's current state of affairs is problematic and unacceptable, then he or she needs to be helped to picture a new state of affairs, that is, alternate, more acceptable possibilities. For instance, for a couple whose marriage is coming apart and who fight constantly, a better future might include greater mutual respect, more openness, fewer and fairer fights, a more equitable distribution of household tasks, and so forth. Separation or even divorce might be considered if differences are irreconcilable and if the couple's values system permits such a solution.

II-B. "What do you really want?" Help clients translate preferred-scenario possibilities into goals. Once a variety of possibilities have been generated, it is time to help clients choose the possibilities that make most sense for them and turn them into an agenda, that is, a goal or a set of goals that form the first part of a constructive change program. A better future is not a wild-eyed, idealistic state of affairs, but rather a picture of what the problem situation would be like if improvements were made. The agenda put together by the client needs to be viable, that is, capable of being translated into action. It is viable to the degree that it is stated in terms of clear and specific outcomes and is substantive or adequate, realistic, in keeping with the client's values, and capable of being accomplished within a reasonable time frame.

II-C. "What are you willing to pay for what you want?" Help clients commit themselves to the goals they choose. Problem-managing goals are useless if they are not actively pursued by clients. They need to determine what they are willing to pay for a better future and then search for incentives to help them move forward. The search for incentives is especially important when the choices are hard. How are truants with poor home situations to commit themselves to returning to school? What are the incentives—the payoff, if you will—for such a choice? Most clients, like most of the rest of us, struggle with commitment to constructive change.

STAGE III: "HOW DO I GET WHAT I NEED AND WANT?"
HELPING CLIENTS DEVELOP STRATEGIES
FOR ACCOMPLISHING GOALS

Discussing and evaluating possibilities for a better future and choosing goals—the work of Stage II—determines the **"what"** of the helping process, that is **what** must be accomplished by clients in order to manage their lives more effectively. Stage III deals with **how** these goals are to be accomplished. Some clients know what they want to accomplish, but need help in determining how to do it.

III-A. Help clients brainstorm a range of strategies for accomplishing their goals. In this step counselors help clients discover a number of different way of achieving their goals. The principle is simple. Action strategies tend to be more effective when chosen from among a number of possibilities. Some clients, when they decide what they want, leap into action, implementing the first strategy that comes to mind. While such a bias toward action may be laudable, the strategy may be ineffective, inefficient, uncreative, imprudent, or a combination of all of these.

III-B. Help clients choose action strategies that best fit their needs and resources. If you do a good job in the first step of Stage III, that is, if you help clients identify a number of different ways of achieving their goals, then clients will face the task of choosing the best set. In this step your job is to help them choose the strategy or "package" of strategies that best fits their preferences and resources. This tailoring of action strategies is important. One client might want to improve her interpersonal skills by taking a course at a college while another might prefer to work individually with a counselor while a third might want to work on his own.

III-C. Help clients draw up a plan. Once clients are helped to choose strategies that best fit their styles, resources, and environments, they need to assemble these strategies into a **plan**, a step-by-step process for accomplishing a goal. If a client has a number of goals, then the plan indicates the order in which they are to be pursued. Clients are more likely to act if they know what they need to do and in what order they do it. Plans help clients develop discipline and also keep them from being overwhelmed by the work they need to do.

THE ACTION ARROW—"HOW DO I GET RESULTS?"
HELPING CLIENTS IMPLEMENT
THEIR PLANS

Helping is ultimately about problem-managing and opportunity-developing **action** that leads to outcomes that have a positive impact on clients' lives. Discussions, analysis, goal setting, strategy formulation, and planning all make sense only to the degree that they help clients to act prudently and accomplish problem-managing and opportunity-developing goals. There is nothing magic about change; it is hard work. If clients do not act on their own behalf, nothing happens.

Two kinds of client action are important here. First, actions within the counseling sessions themselves. The nine steps described above are not things that helpers do to clients, rather they are things that clients are helped to do. Clients must act by taking ownership of the helping process. Second, clients must act "out there" in their real day-to-day worlds. Problem-managing and opportunity-developing action is ultimately the name of the game.

Since all the stages and steps of the helping process can be drivers of client action, a number of exercises in this manual will focus on action themes. This will reinforce the principle that discussion and action must go hand in hand.

The next two exercises continue to help you prepare yourself to play, as realistically as possible, the role of client in the training group. These two sentence-completion exercises focus on both problems and opportunities.

EXERCISE 7: PREPARING TO TELL YOUR STORY—PROBLEMS

All helpers have problems or issues that could lessen their effectiveness. Knowing what yours are is an important part of training. This and the following sentence-completion exercise will help you to choose issues that you would like and be willing to discuss. Do these sentence-completion exercises quickly.

1. One of my biggest problems is

2. I'm quite concerned about

3. Something I fail to do that gets me into trouble is

4. The social setting of life I find most troublesome is

5. The most frequent negative feelings in my life are

6. These negative feelings take place when

7. The person I have most trouble with is

8. What I find most troublesome in this relationship is

9. Life would be better if

10. I tend to shortchange myself when I

11. I don't cope very well with

12. I get anxious when

13. A value I fail to put into practice is

14. I'm afraid to

15. I wish I

16. I wish I didn't

17. What others dislike most about me is

18. I don't seem to have the skills I need in order to

19. A problem that keeps coming back is

20. If I could change just one thing in myself it would be

Once you have finished, check the issues that you might be willing to discuss in the training groups. Then circle the checks you believe are most closely related to becoming an effective helper. Finally, with a learning partner discuss the items you have both checked and circled.

EXERCISE 8: PREPARING TO TELL YOUR STORY—UNUSED OPPORTUNITIES

This sentence-completion exercise mirrors the preceding one but deals with unused or underused resources and opportunities rather than problems. Undeveloped resources and opportunities can affect the quality of your helping.

1. One thing I like about myself is

2. One thing others like about me is

3. One thing I do very well is

4. A recent problem I've handled very well is

5. When I'm at my best I

6. I'm glad that I am able to

7. Those who know me know that I can

8. A value that I try hard to practice is

9. An example of my caring about others is

10. People can count on me to

11. Something I'm handling better this year than last is

12. A good example of my ability to manage my life is

13. I'm at my best when

14. One goal I'm presently working toward is

15. A recent temptation that I managed to overcome was

16. I think that I have the guts to

17. If I had to say one good thing about myself, I'd say that I

18. One way I successfully control my emotions is

19. One way in which I am very dependable is

20. I communicate most effectively with others when

Once you have finished, check the unused opportunities that you might be willing to discuss in the training groups. Then circle the checks you believe are most closely related to becoming an effective helper. Finally, discuss the items that you have both checked and circled with a learning partner.

Chapter 3

THE HELPING RELATIONSHIP:
VALUES IN ACTION

It is important for you to take initiative in determining the kinds of values you want to pervade the helping process and relationship. Too often such values are afterthoughts. The position taken in Chapter Three of *The Skilled Helper* is that values should provide guidance for everything you do in your interactions with clients.

As indicated in the text, the values that are to permeate your helping relationships and practice must be owned by you. Learning about values from others is very important, but mindless adoption of the values promoted by the helping profession without reflection inhibits your owning and practicing them. You need to wrestle with your values a bit to make them your own. Second, your values must actually make a difference in your interactions with clients. That is, they must be values-in-use and not merely espoused values. Your clients will get a feeling for your values, not from what you say, but from what you do.

Do the following exercises BEFORE reading Chapter Three.

EXERCISE 9: WHAT IF I WERE A CLIENT?

1.   Picture yourself as a client. Think of some of the problems you have had to grapple with or are struggling with now. Then picture yourself dealing with these with a counselor. Finally, ask yourself the following questions. Jot down words, phrases, or simple sentences in response.

What would I want the helper to be like?

_____

_____

_____

_____

_____

_____

How would I want to be treated?

_____

_____

_____

_____

_____

2.   Share your statements with a learning partner. Discuss both the similarities and the differences between your responses.

EXERCISE 10: YOUR PRELIMINARY STATEMENT OF VALUES

1.   On a separate piece of paper write a preliminary STATEMENT OF VALUES based on what you learned from the previous exercise. Write it in such a way that you could give it to a client, saying, "These are the values that underlie and permeate the helping process."

2.   Share your statement with a learning partner. Note both the similarities and the differences. Tell each other how you would feel, in light of the other's statement of values, what your hopes and fears would be were you to be your learning partner's client.

3.   Finally, read Chapter Three in *The Skilled Helper*. That chapter deals with values that should permeate the helping process together with a preliminary consideration of the relationship between client and helper. In the light of what you learn, revise your value statement and share it with a learning partner.

EXERCISE 11: BUILDING AND MAINTAINING RELATIONSHIPS

It goes without saying that the relationship between helper and client is important. Your job is to build the kind of relationship with each client that contributes to problem-managing and opportunity-developing outcomes. This exercise asks you to look at your strengths and weakness in establishing and maintaining relationships in your everyday life.

Briefly describe your overall effectiveness in establishing solid working relationships with other people.

_____

_____

What do you do best in establishing and maintaining relationships?

_____

_____

What mistakes do you make?

_____

_____

What do you do to recover from mistakes?

_____

_____

Discuss your responses with a learning partner. What do you need to do to become better at establishing and maintaining good working relationships with others?

_____

_____

PART TWO

THE THERAPEUTIC DIALOGUE

In Part Two of *The Skilled Helper,* the basic communication skills needed to be an effective helper are reviewed and illustrated. These skills play a critical role in your "therapeutic dialogue" with clients. The therapeutic dialogue is, of course, the helping dialogue, the problem-management dialogue, the opportunity-development dialogue. Dialogue emphasizes the two-way nature of helping. Dialogue is at the heart of the communication between helper and client. Chapter 4 includes an overview of both interpersonal communication and dialogue together with the first basic skill, often called "attending," but now called, more pragmatically, "visibly tuning in" to clients. This skill focuses on your empathic presence to the client. Chapter 5 focuses on the skill of active listening. You visibly tune in, not only to demonstrate your solidarity with your clients, but also to understand what your clients are saying both directly and indirectly. Finally, Chapter 6 deals with the skill you need to both check out and share your understanding with clients. This skill, called basic empathy in previous editions, has been renamed "sharing empathic highlights" to distinguish it from empathy as a value which should permeate all helping skills.

Chapter 4

COMMUNICATION SKILLS I:
DIALOGUE AND VISIBLY TUNING IN

Much of helping takes place through a dialogue between client and helper. If this dialogue is to serve the overall problem-management and opportunity-development goals of the helping process, the quality of the dialogue is critical.

THE IMPORTANCE OF DIALOGUE

As useful as dialogue might be in human communication, it is not that common—at least in its fullness.

EXERCISE 12: UNDERSTANDING THE FOUR REQUIREMENTS OF TRUE DIALOGUE

1. Read the section in Chapter Four that describes the requirements of a true dialogue. This describes what dialogue should look like.
2. Describe what happens in a conversation when each of these requirements is missing.
3. Compare your responses with those of a learning partner.

Turn-Taking

_____

_____

_____

Connecting

_____

_____

_____

Mutual Influence

_____

_____

_____

Co-Creating Outcomes

_____

_____

_____

EXERCISE 13: DIALOGUE IN EVERYDAY LIFE

Not all conversations need be true dialogues. However, most conversations would probably be more productive if they were conducted in the *spirit* of dialogue. And, often enough, there are times when dialogue in its fullest sense would make a big difference. In this exercise you are asked to spend the next week or two watching yourself and others in conversations. See how much time you and your conversational partners spend in dialogue, especially when important topics are being considered.

1.   First rate yourself. On a scale from 1-10, how good are you at dialogue in everyday life? _____
2.   Then watch yourself and your conversational partners in conversations.
3.   How well did you do in dialogue, especially when true dialogue would have added value to the conversation?
4.   At the end of this experiment, re-rate yourself on dialogue. What's your new score? _____
5.   What could you do to improve your ability to engage in dialogue?

_____

_____

_____

_____

ATTENDING: VISIBLY TUNING IN

Your posture, gestures, facial expressions, and voice all send nonverbal messages to your clients. The purpose of the exercises in this section is to make you aware of the different kinds of nonverbal messages you send to clients through such things as body posture, facial expressions, and voice quality and how to use nonverbal behavior to make contact and communicate with them. It is important that what you say verbally is reinforced rather than muddled or contradicted by your nonverbal messages. Before doing these exercises, read the rest of Chapter Four in *The Skilled Helper*. There are two messages in that chapter. First, use your posture, gestures, facial expressions, and voice to *send messages* you want clients to hear, such as, "I'm listening to you very carefully" or "I know what you're saying is difficult for you, but I'm with you." Second, become aware of the messages your clients are sending to you through their nonverbal behaviors.

EXERCISE 14: VISIBLY TUNING IN TO OTHERS IN EVERYDAY CONVERSATIONS

This is another exercise you do outside the training group in your everyday life. Observe the way you attend to others for a week or two—at home, with friends, at school, at work. Observe the quality of your presence to others when you engage in conversations with them. Of course, even being asked to "watch yourself" will induce changes in your behavior; you will probably "tune in" more effectively than you ordinarily do. The purpose of this exercise is to sensitize you to attending behaviors in general and to get some idea of what your day-to-day attending style looks like. First, read about the skills of tuning in or attending in *The Skilled Helper*.

1. On a scale from 1-10, rate yourself on how effectively you think you pay attention or tune in to others in your everyday conversations. _____

2. Without becoming preoccupied with every little behavior, watch yourself for a week as you tune in (or fail to tune in) to others in your everyday conversations.

3. What are you like when you are with others, especially in serious situations? What do you do well?

4. Re-rate yourself _____

5. How can you improve the ways you are present to others in conversations?

_____

_____

_____

_____

_____

_____

Example. Here is what one trainee wrote: "I found myself tuning in better to people I like. When I was with someone neutral, I found that my eyes and my mind would wander. I also found that it's easier for me to tune in to others when I'm rested and alert. When I'm physically uncomfortable or tired, I don't put in much effort to tune in. I was unpleasantly surprised to find out how easily distracted I am. Often I was there, but I really wasn't there. However, simply by paying attention to tuning in skills, I was with others more fully, even with neutral people."

6. Write a summary of what you have learned about yourself.

_____

_____

_____

_____

_____

7. What can you do to tune in more effectively to others?

_____

_____

_____

_____

_____

EXERCISE 15: OBSERVING AND GIVING FEEDBACK ON QUALITY OF PRESENCE

In the training sessions, make sure that your nonverbal behavior is helping you work effectively with others and send the messages you want to send. This exercise, then, pertains to the entire length of the training program. You are asked to give ongoing feedback both to yourself and to the other members of the training group on the quality of your presence to one another as you interact, learn, and practice helping skills. Recall especially the basic elements of visibly tuning in summarized by the acronym *SOLER*. Here is a checklist to help you provide that feedback to yourself and your fellow helpers. Review criteria for giving useful feedback found in the Introduction to the Exercises.

- How effectively are you using postural cues to indicate your willingness to work with the client?
- In what ways do you distract clients and observers from the task at hand, for instance, by fidgeting?
- How flexible are you when engaging in *SOLER* behaviors? To what degree do these behaviors help you be with the client?
- How natural is it for you to tune in to the client? What indications are there that you are not being yourself?
- To what degree is your psychological presence reflected in your physical presence?
- What are you like when you miss the mark?
- What are you like when you are at your best?

What do you have to do to become more effectively present to your clients both physically and psychologically?

fidgety ie...picking at face, ears
Stay grounded in my recovery, physically eating right, exercise
mentally stable – sharing feelings – life
spiritual fitness – grounded.

Of course, more important than nonverbal behavior in itself is the total quality of your being with and working with your clients. Your posture and nonverbal behavior are a part of your presence, but there is more to presence than *SOLER* activities. There are the values and spirit you bring to your encounters with your clients.

Chapter 5

ACTIVE LISTENING

Read the sections on listening in Chapter Five of *The Skilled Helper*. Effective helpers are active listeners. When you listen to clients, you listen to:

- their **stories**. These stories include:
  - their **experiences**, what they see as happening *to* them;
  - their **behaviors**, what they do or fail to do;
  - their **affect**, the feelings and emotions that arise from their experiences and behaviors;
  - the **core messages** in their stories;
- their **points of view** about key topics, including the reasons for their points of view and the implications for holding any given point of view;
- the **decisions** they are making together with the reasons for them and the implications or possible consequences of their decisions;
- their **intentions**, that is, the goals they are pursuing and the actions they intend to engage in;
- the wider **context** of their stories, points of view, decisions, and intentions;
- the **slant** they might give to all of the above.

Helpers listen carefully in order to respond with both understanding and, as we shall see later, some sort of challenge. Let's start by having you listen to yourself.

EXERCISE 16: LISTENING TO YOURSELF AS PROBLEM SOLVER

In this exercise you are asked to "listen to" some problem situation or unused opportunity you have dealt with successfully. Retell the story to yourself in summary form.

What was the issue?

Cancellation of mental health day

What were your key experiences, that is, what happened to you?

I felt angry, distracted from my work, unstable, rebellion —.

What actions on your part contributed to the problem situation?

Writing / sending e-mail to Headquarter

What points of view of yours were involved?

thought decision unfair, inappropriately handled.

What decisions were involved?

① Deciding to send e-mail to Bishop

② Taking leave

What emotions did you experience?

frustration, anger, hurt,

What did you do to cope successfully with the problem or develop the opportunity?

admitted my part — breaking policy realizing how emotional I was because of little Teal's funeral the day before. talking w/ law about no regrets knowing I broke policy, and accepting possible consequences

Share what you have written with a learning partner. Discuss how well both of you have listened to yourselves in outlining the core messages and issues of the problem or opportunity.

EXERCISE 17: LISTENING TO YOUR OWN FEELINGS AND EMOTIONS

Since emotions are strong motivators for good or for bad and permeate stories, points of view, decisions, and intentions, we turn to an exercise on emotions. If you are to listen to the feelings and emotions of clients, you first should be familiar with your own emotional states. A number of emotional states are listed below. You are asked to describe what you feel when you feel these emotions. Describe what you feel as *concretely* as possible: How does your body react? What happens inside you? What do you feel like doing? Consider the following examples.

Example 1 — Accepted:

When I feel accepted,

- I feel warm inside.
- I feel safe.
- I feel free to be myself.

Example 2 — Scared:

When I feel scared,

- my mouth dries up.
- my bowels become loose.
- there are butterflies in my stomach.
- I feel like sitting back and relaxing.

- I feel I can let my guard down.
- I feel like sharing myself.
- I feel some of my fears easing away.
- I feel at home.
- I feel at peace.
- I feel my loneliness melting away.

- I feel very uncomfortable.
- I feel the need to talk to someone.
- I turn in on myself.
- I'm unable to concentrate.
- I feel very vulnerable.
- I sometimes feel like crying.
- I feel like running away.

1. Choose four of the emotions listed below or others not on the list. Try your hand at the emotions you have difficulty with. It's important to listen to yourself when you are experiencing emotions that are not easy for you to handle.
2. Picture to yourself situations in which you have actually experienced each of these emotions.
3. Then, as in the example above, write down in detail on a separate sheet of paper what you experienced.

1. accepted	12. disappointed	23. lonely
2. affectionate	13. free	24. loving
3. afraid	14. frustrated	25. rejected
4. angry	15. guilty	26. respected
5. anxious	16. hopeful	27. sad
6. attracted	17. hurt	28. satisfied
7. bored	18. inferior	29. shocked
8. competitive	19. interested	30. shy
9. confused	20. intimate	31. superior
10. defensive	21. jealous	32. suspicious
11. desperate	22. joyful	33. trusting

The reason for this exercise is to sensitize yourself to the wide variety of ways in which clients express and name their feelings and emotions.

Although the feelings and emotions of clients (not to mention your own) are extremely important, sometimes helpers concentrate too much, or rather too exclusively, on them. Feelings and emotions need to be understood, both by helpers and by clients, in the **context** of the experiences and behaviors that give rise to them. On the other hand, when clients hide their feelings, both from themselves and from others, then it is necessary to listen carefully to cues indicating the existence of suppressed, ignored, or unmanaged emotion.

LISTENING THOUGHTFULLY TO CLIENTS' STORIES

Listening thoughtfully means identifying the key elements of clients stories—experiences, behaviors, feelings—and the relationships among them. Thoughtful listening is a function of empathy.

EXERCISE 18: PROCESSING WHAT YOU HEAR—THOUGHTFUL LISTENING

Listening in a thoughtful way goes beyond being able to recall what the client has said. It means being able to discriminate between the details of the client's story and the really meaningful points—the core messages. This exercise is intended to help you identify core messages.

1. Read the section in the book about "Processing What You Hear: The Thoughtful Search For Meaning." Pay particular attention to the extended example of Denise and Jennie.

2. Share a problem or opportunity from your own life with your learning partner. To provide a good learning experience you and your partner should spend some five minutes in dialogue.
3. When you have finished, both you and you partner should write down separately three essential elements of the story you have related—the core messages.
4. Reverse roles and give your learning partner a chance to do the same thing.
5. Compare notes with one another.
    - How well, in the eyes of your partner, did you capture his or her core messages?
    - What helped you separate the core messages from the rest of the story?
    - What difficulty do you have in capturing core messages?
6. What have you learned about yourself?

_____

_____

_____

EXERCISE 19: LISTENING FOR CORE MESSAGES

Core messages are the main points of a client's story. The ingredients of core messages are key experiences and key behaviors together with the key feelings or emotions associated with them. In this exercise you are asked to "listen to" and identify the key experiences and behaviors that give rise to the client's main feelings.

1. Listen very carefully to what the client is saying.
2. Identify the client's key experiences, what he or she says is happening to him or her.
3. Identify the client's key behaviors, what he or she is doing, not doing, or failing to do.
4. Identify the key feelings and emotions associated with these experiences and behaviors.

Example: A twenty-seven-year-old man is talking to a minister about a visit with his mother the previous day. He says, "I just don't know what got into me! She kept nagging me the way she always does, asking me why I don't visit her more often. As she went on, I got more and more angry. (He looks away from the counselor down toward the floor.) I finally began screaming at her. I told her to get off my case. (He puts his hands over his face.) I can't believe what I did. I called her a bitch. (Shaking his head.) I called her a bitch several times and then I left and slammed the door in her face."

Key experiences: Mother's nagging.
Key behaviors: Losing his temper, yelling at her, calling her a name, slamming the door in her face.
Feelings/emotions generated: He now feels embarrassed, guilty, ashamed, distraught, disappointed with himself, remorseful. (Note carefully: This man is not at this moment **expressing** anger. Rather he is talking **about** his anger, the way he let his temper get away from him.)

Now do the same with the following cases.
1. A woman, 40, married with no children, has had several sessions with a counselor. She went because she was bored and felt that all the color had gone out of her life. In a later session she says this: "These counseling sessions have really done me a great deal of good! I've worked hard in these sessions, and it's paid off. I enjoy my work more. I actually look forward to meeting new people. My husband and I are talking more seriously and decently to each other. At times he's even tender toward me the way he used to

be. Now that I've begun to take charge of myself more and more, there's just so much more freedom in my life!"

Client's key experiences:

bored with life no excitement (color)

Client's key behaviors:

willing to accept help and work on herself

How does the client feel about these experiences and behaviors?

↑ self esteen   hopeful
satified   encouraged

2. A college student, 20, who has volunteered to work with struggling high school students finds that he is learning more than he imagined he would. He tells his college advisor what it's like: "These kids deal with problems I have never even had a brush with in my life. Some days we'll start by working on some bit of homework but before we're done they're talking about being afraid of the gangs on the way to and from school. This one kid starts and ends each day by helping his mom who has MS. In the morning he gets her up and dresses her and in the evening he gets her ready for bed. Some days I do as much listening as talking. These kids have a type of maturity about them you don't see in people who have lived comfortably."

Client's key experiences:

working / teaching kids who want to talk about their lives

Client's key behaviors:

listening to his kids,

How does the client feel about these experiences and behaviors?

excited, shocked, hopefull

3. A woman, 38, unmarried, talking about a situation with a friend: "My best friend has just turned her back on me. And I don't even know why! (said with great emphasis) From the way she acted, I think she has the idea that I've been talking behind her back. I simply have not! (also said with great emphasis) Damn! This neighborhood is full of spiteful gossips. She should know that. If she's been listening to those foul-mouthed jerks who just want to stir up trouble. . . . She could at least tell me what's going on."

Client's key experiences:

rejection from a friend

Client's key behaviors:

defensiveness, frustration

How does the client feel about these experiences and behaviors?

rejected, angry,

4. A man, 64, who has been told that he has terminal lung cancer, is speaking to a medical resident: "Why me? I'm not even that old! I keep looking for answers and there are none. I've sat for hours in church and I come away feeling empty. Why me? I don't smoke or anything like that. (He begins to cry.) Look at me. I thought I had some guts. I'm just a mess. Oh God, why terminal? What are these next months going to be like? (Pause, he stops crying.) Why would you care! I'm just a failure to you guys."

Client's key experiences:

found out he has terminal cancer

Client's key behaviors:

venting w/ need. resolution

crying

sharing feeling

How does the client feel about these experiences and behaviors?

ashamed for crying

hopeless, afraid

5. A woman, 54, talking to a counselor about a situation at work: "I don't know where to turn. They're asking me to do things at work that I just don't think are right. If I don't do them—well, I'll probably be let go. And I don't know where I'm going to get another job at my age in this economy. But if I do what they want me to, I think I could get into trouble, I mean legal trouble. I'd be the fall guy. My head's spinning. I've never had to face anything this before. . . . Where do I turn?"

Client's key experiences:

Client's key behaviors:

How does the client feel about these experiences and behaviors?

6. A bachelor, 39, is speaking to the members of a group in which the members meet bi-weekly to discuss important issues in their lives. They call it a "lifestyle" group. He has belonged to this group for about a year: "I've talked a lot about my rather meager social life. But I've finally met a woman who is very genuine and who lets me be myself. We seem to be able to care deeply about each other without feeling dependent on each other. (He is speaking in a soft, steady voice.) I've always had a horror of dependency—going either direction. And she cares about me without mothering me. I never thought it would happen. (He raises his voice a bit.) Is this actually happening to me? Is it actually happening?"

Client's key experiences:

_____

_____

_____

Client's key behaviors:

_____

_____

_____

How does the client feel about these experiences and behaviors?

_____

_____

_____

LISTENING TO CLIENTS' POINTS OF VIEW, DECISIONS, AND INTENTIONS

Interweaved with clients' stories, goal setting, and plans are their points of view, decisions, and intentions. These need to be listened to carefully.

EXERCISE 20 LISTENING FOR REASONS, IMPLICATIONS, AND FEELINGS

1. A husband and wife have entered counseling to help them deal more effectively with their daughter. The daughter is fifteen and is struggling in school and becoming involved with older boys. At this point the husband offers his point of view: "It seems to me that we both need to stay on top of her school work. We need to do a better job of that—no question. This business with the boys, though, I think should be handled primarily by her mother. I'll back up whatever needs to be said but there is a part of this which is best handled between two women."

What is his point of view?

women should deal w/ girl's issues, he can assist / help with homework

What are the reasons behind it?

think men shouldn't or are unable to deal w/ female issues

What are some possible implications of this point of view?

fear, family history

31

2. A single parent talks to a school counselor about the fact that his son is being picked on by others in his class. He has discussed the facts of the matter and feels the counselor has a good grasp of what is going on. Not surprisingly he has a strong point of view as well. "If nothing is done, I'll have to take matters into my own hands and talk with the parents of the other boys. But that's not what I would prefer. I see this as a school matter and I will work with you and the teachers to see if this can be resolved in a sensible way. I know 'boys will be boys,' but this has gone too far already."

What is the point of view?

School should deal w/ school issue, parent are responsible as last resort.

What are the reasons behind it?

possibly avoidance father may possibly be afraid to confront situation

What are some possible implications of this point of view?

fear

3. A manager of a project team has sought out the Director of Human Resources to discuss a matter of discipline. She uses the HR director to sound out a decision he is making. "I've decided I need to put Clarence on probation. I've wanted to avoid this but he's given me no choice. The performance improvement plan he's been on the last two months has been a bust. He still comes late or calls in sick with no notice. We're short-handed as it is and when he does this stuff, it nearly brings the operation to a halt. The others on his shift are up in arms and I've already said this is the next step."

What is her decision?

What are the reasons behind it?

What are some possible implications of this decision?

_____

_____

_____

4. An unmarried woman, 33, has recently learned that she is HIV-positive. "When I first learned about this I was devastated. The end. The curtain is coming down. But one morning I woke up and said to myself, 'Do something.' I did. I got on the net and learned everything I could about this. I spent days studying it. And, you know, I'm going to beat it. I'm going to get physically fit. I'm going to get my head together. I'm going to take the drugs that keep you going. I'm not sure how I'm going to afford them, but I'll get them. I'm going to volunteer for some experimental approaches. I don't know how you get on the list, but somehow I will. I'm going to beat this thing."

What is she intending to do?

_____

_____

How is she going to make it work?

_____

_____

_____

How does she feel about all of this?

_____

_____

5. A man, 47, works in the construction trade and has been experiencing an increasing number of days when his knees and back ache to the point of hardly being able to get dressed in the morning. He is talking with a friend and floats this possibility: "I don't know what other work I can get into at this point in my life but I can't go on like this much longer. Ten years ago I'd have been made a supervisor and been able to hang in there but those days are gone. I'm taking painkillers every day and in the last year my right knee especially has been . . . well, I can't continue. What if I just made the move? I know I could do real estate appraisal. It might start slow but I think it would work."

What does he intend to do?

_____

_____

What are the reasons behind it?

_____

_____

_____

What are some of his concerns about what he intends to do?

_____

_____

_____

6. A woman, 55, is talking to her brother about the care their aging mother requires. She makes a case for a new arrangement which she thinks will benefit their mother: "Mom likes being home and I think we can make that happen for her but it will take some doing. We already have Meals on Wheels helping out. What I think we'll need to do is work out an arrangement where you and I handle the weekends and we get a live-in companion to be there Monday through Friday. There are a lot of details to work out, but what do you think? And it's going to cost us some money."

What is the heart of the proposal?

_____

_____

What are the reasons behind it?

_____

_____

_____

What are some possible implications of the proposal?

_____

_____

_____

LISTENING THOUGHTFULLY TO OPPORTUNITIES

Spotting and understanding clients' opportunities is just as important as understanding their problems. Often exploiting opportunities is the best way to manage problems.

EXERCISE 21: LISTENING TO KEY OPPORTUNITIES

The instructions for this exercise are the same as those for the previous exercise.

1. Ken, 45, has just been passed up for a promotion. He is talking with his supervisor who was a strong supporter. He starts by talking about his disappointment. But his supervisor points out that staying in his present position will allow him to see a project he started through to completion. Ken shifts his focus to the opportunity side of the event: "You may have a point. In fact, during my entire career I've been so focused on my next job that I've tended to do a lot of switching horses in mid-stream—taking new jobs before projects were really completed. Don't get me wrong—I'm not giving up on seeking promotion around here. But, I like what I'm doing and now I can give it my undivided energy. I suppose that people

have to see that I can see things through, that I can get results with the best of them. Moving too quickly in the past has prevented them from seeing me at my best."

What is the opportunity?

_____

_____

What's the up side and the down side?

_____

_____

_____

How does Ken feel about it?

_____

_____

2.   A woman, 43, has suffered a major financial setback in her small consulting business. She has just been offered a good position in a large consulting firm. "Lots of people would jump at this offer. The firm is solid, the job is challenging, and the pay is good. I'm really tempted. But I have really enjoyed being my own boss and building something out of nothing. If I take the offer, I'll have a boss again. Though I suppose that the economy has been my boss. The downturn is killing me. The little guys are the first to lose business. I do like the security I would have in the new job. But I know I won't be able to be as creative as I am when I'm on my own."

What is the opportunity?

_____

_____

What's the up side and the down side?

_____

_____

_____

How does she feel about it?

_____

_____

UNDERSTANDING CLIENTS' PROBLEMS AND OPPORTUNITIES THROUGH CONTEXT

EXERCISE 22: USING CONTEXT FOR THOUGHTFUL PROCESSING

Becoming a thoughtful processor requires two things: following the main points of the client's story, point of view, decision, or intention as it unfolds *and* at the same time placing it in the context of their lives. Since we tend to know the context of our own lives, this exercise makes the point about thoughtful processing by asking you to focus on yourself.

1.   Read the section in the book about "Processing What You Hear: The Thoughtful Search for Meaning."
2.   Share a problem or unexploited opportunity from your own life with your learning partner. **Provide no context**.
3.   Discuss the issue from five to ten minutes.
4.   When you have finished, tell your partner about the unique circumstances of your life that give added meaning to what you have shared. Discuss how the context changes the meaning of the problem or unexploited opportunity.
5.   Reverse roles and give your learning partner a chance to do the same thing.

Example. A trainee, 24, discusses how difficult it is to keep up with the demands of her program while parenting a one-year-old child. Her husband works long hours and, though he helps where he can, much of the childcare burden falls to her.

Context. The trainee later provides the following background: "I am a capable woman who is happy to be a mother but I find this at odds with my professional aspirations and my feminist beliefs. I have a strong desire to become more a part of the training program but my responsibilities at home prevent me from doing this. I am very social by nature and I miss out on the fun others in the program have when they go out together after class. My strengths are many. I'm deeply committed to my marriage, my child, my family, and my many friends. I'm in good health. I'm open about my frustrations; I don't let them simply stew inside me. My husband is in an early stage of his career. He faces his own set of challenges, many of which consume a great deal of his time."

THE SHADOW SIDE OF LISTENING

Since effective helpers are active listeners, it helps to learn about your own listening style—both the up side and the down side.

EXERCISE 23: YOUR LISTENING AND PROCESSING STYLE AND ITS SHADOW SIDE

In this exercise you are asked to take an objective look at your style of listening and the way you process what you hear.

1. What kind of listener are you? Give a brief, balanced description.

_____

_____

_____

_____

_____

_____

2. Re-read "The Shadow Side of Listening to Clients." Indicate below any areas of listening you think you should work on.

_____

_____

_____

_____

_____

_____

3. Discuss your description of yourself as a listener and processor together with your shadow-side self-assessment with a learning partner. Talk about the implications of your shadow-side traits. Indicate possible actions you can take to overcome whatever deficiencies you discover.

Chapter 6

SEEKING UNDERSTANDING
SHARING EMPATHIC HIGHLIGHTS

The payoff of attending and listening lies in the ability to communicate to clients an understanding of their stories, points of view, decisions, and intentions together with the feelings and emotions they generate. Sharing empathic highlights is one way of communicating your understanding. It is one important way in which you translate the value of empathy into behavior. It is a skill you need throughout the helping model. Sharing highlights is critical for establishing and developing relationships with clients and for helping them clarify both problem situations and unexploited opportunities, setting goals, and developing strategies and plans.

EXERCISE 24: IDENTIFYING CORE MESSAGES IN CLIENTS' STORIES

In this exercise, you are asked to identify key experiences, behaviors, and feelings and then translate them into a core message. The core message is a highlight that can be communicated to the client.

1.   Listen carefully to the client's statement.
2.   Identify key experiences, behaviors, and the feelings they generate.
3.   Pull them together in a statement of the client's core message.
4.   After finishing, compare your responses with those of your learning partner. Then see if the two of you can improve upon your combined responses.

Example: Roula is recovering from a bad car accident in a rehabilitation unit. She is a week or two into a program that could well go on for a few months. She is talking with one of the rehabilitation specialists. She says, "You told me it would be tough going. And I thought that I had prepared myself for it. I thought I had some courage in me. But now I can't find any grit at all. Doing even the smallest things takes so much effort! I keep breaking down and crying when I'm alone. I just keep giving in . . . I'm just inches down a path that seems miles long. It seems endless."

> **Key experience(s):** finding the rehabilitation program so difficult
> **Key behavior(s):** inability to tap into the courage she thought she had; giving in to discouragement
> **Key feelings/emotions:** disappointment, discouragement, despair

Core message: Roula seems disappointed almost to the point of despair because she has failed to tap into the "stuff" within her that would keep her going during a very tough rehab program.

1. This woman is about to go to her daughter's graduation from college. She is talking with one of the ministers of her church: "I never thought that my daughter would make it through. I've invested a lot of money in her education. It meant scrimping and saving and not having some of the things I wanted. But money is certainly not the issue. More to the point, I've put a lot of emotion into making this day happen. I had to do a lot of handholding to help her get through. There were times when neither of us thought this day would come. But the day has arrived!"

Key experience(s):

_____

_____

Key behavior(s):

_____

_____

Key feelings/emotions:

pride, relief, guilt
_____

_____

Core message:

relief about daughter graduating
doubt turned into belief
_____

_____

2. This older man has just had his wallet stolen. He became disoriented and was taken to a hospital. He's talking with a social worker. "I had just cashed my paycheck and the money was in the wallet. I've had a streak of bad luck. My sister was in an auto accident last week. And I was on my way to visit my nephew. Ironically, he was detained by the police for a minor theft earlier this week. Now he's probably thinking I've given up on him, when I haven't. There has not been much good news at all for a while."

Key experience(s):

wallet was stolen, dealing w/ grief
issues, bad experience
_____

Key behavior(s):

_____

_____

Key feelings/emotions:

_____

_____

Core message:

depressed about current life circumstances
not sure where his life is happening / what is
going
_____

3. This woman is waiting for the results of medical tests. She is talking to a hospital volunteer: "I've been losing weight for about two months and feeling tired and listless all the time. I'm afraid of what these tests are going to say. I know I've been putting them off. I just hate hospitals and this kind of stuff. Well, now the waiting is getting to me. I . . . well, I just don't know where I stand. Nobody said anything to me during the tests. I don't think that's a good sign. It's all so impersonal anyway."

Key experience(s):

_____

_____

Key behavior(s):

_____

_____

Key feelings/emotions:

_____

_____

Core message:

_____

_____

_____

4. A prospective employer has just found out that this client has a criminal record: "I had hoped that I would get the job and prove myself before anyone found out about my record. I guess I was just stupid. I've just received a call from him telling me that I'm no longer being considered for the job. Well, I did what I thought was right. I never had the intention of deceiving anyone. I mean that I didn't think that I was doing anything wrong . . . . Well, even though I'm branded because of my record, I'm going to make it, somehow."

Key experience(s):

_____

_____

Key behavior(s):

_____

_____

Key feelings/emotions:

_____

_____

Core message:

_____

_____

_____

5. This woman has just lost a custody case for her only son. During the interview with a counselor she seems almost in a daze: "I never dreamed that the court would award custody to my husband. I may not be a perfect mother, but I've done all the work in raising him so far. He's done nothing but give me a hard time over the years. He's so selfish and spiteful. I've been racking my brains trying to come up with something I can do . . . . Now he's won . . . . It's all over."

Key experience(s):

_____

_____

Key behavior(s):

_____

_____

Key feelings/emotions:

_____

_____

Core message:

_____

_____

_____

EXERCISE 25: USING A FORMULA TO EXPRESS EMPATHIC HIGHLIGHTS

Empathic highlights focus on the client's key messages—key experiences and/or key behaviors, key points of view, key decisions, key intentions—plus the feelings and emotions they generate. In this exercise you are asked to use the formula explained in the text: "You feel . . . " (followed by the right emotion and some indication of its intensity) "because . . . " (followed by the key experiences and/or behaviors that give rise to the emotion).

1. Identify for yourself the speaker's key experiences, behaviors, points of view, decisions, intentions, and the feelings associated with them.
2. Formulate an empathic response. Even though you are writing the response, picture yourself actually talking to the person.

Example: A woman in a self-help group is talking about a relationship with a man. She says: "About a couple of months ago, he began being abusive, calling me names, describing my defects. To tell you the truth, that's why I joined this group, but I haven't had the courage to talk about it till now. The couple times I've tried to stand up for myself, he became even more abusive. He hasn't hit me or anything, but, anyway. So I've been just taking it, just sitting there taking it . . . like a dog or something. Do you think that this is just his bizarre way of getting rid of me? Why doesn't he just tell me?"

> **Key experience(s)**: being abused by companion, escalation in abuse
> **Key behavior(s)**: trying to stand up for herself, becoming passive, trying to figure out if this is his way of getting rid of her
> **Key feelings/emotions**: distraught, confused, angry

Empathic highlight (using formula): You feel angry and confused because the abuse came from out of the blue and now you're wondering whether this is just his strange way of ending the relationship.

1. First-year college student talking to a counselor in the Center for Student Services. She has been talking about some of the difficulties of adjusting to college life. She comes from a small town and is attending a large state university. She speaks openly and seems to be in good spirits; she even smiles at times: "My friends seem so much more sophisticated than I am. Just the other night Mairead and Johanna showed up at the party in dresses *and* with dates. And there I was, alone and in jeans! I mean, just typical of me! . . . I guess I was more amazed than embarrassed. And I did have a good time anyway. It's like I have to learn a whole new set of rules. But nobody just comes out and says what they are."

Key experience(s):

not feeling part of

Key behavior(s):

Key feelings/emotions:

afraid, rejected

Empathic highlight (formula):

It sounds like your adjusting to being at a large college and you don't feel accepted. *

2. A 66-year-old man is talking to a mental health counselor: "My wife died last year, and this year my youngest son went away to college. The other children are married. So now that I'm retired, I spend a lot of time rambling around a house that's . . . (pauses, look out the window for a while) . . . really too big for me. . . . You know, when I was working, there was a certain fullness to life. I always knew what to do. I, well . . . I made a difference. Now that I've got a comfortable retirement, I . . ."

Key experience(s):

_____

_____

Key behavior(s):

_____

_____

Key feelings/emotions:

_____

_____

Empathic highlight (formula)

_____

_____

_____

3. This 33-year-old woman has been working with a counselor to find ways of handling sex discrimination at work. The company seem to preach one philosophy but implement another. Up to this point she has had some modest success, but now she wants to discuss a setback: "It seems they let me push as long as I stayed with the little stuff and did it quietly. But last week I brought up discrimination in our monthly team meeting . . . and my boss clammed up. And he's not one of the worst offenders. He's been quite cool since then and always seems to be busy. All of a sudden there's quite a different atmosphere in the office."

Key experience(s):
_____ sex descrim̄te at work _____

_____

Key behavior(s):

_____

_____

Key feelings/emotions:

_____

_____

Empathic highlight (formula):
_____ It sounds like your feeling uncomfortable
for being honest with your co-work
and are not sure wht do next.
                              * vulnerable

4. Man, 40, talking about his invalid mother to a minister at his church: "She uses her illness to control me. It's a pattern; she's been controlling me all her life. The times I try to stand up to her, well, she has her way of dealing with that, too. (He grits his teeth and sets his jaw.) I bet she'll even make me feel guilty for her death."

Key experience(s):

Controlling mother

Key behavior(s):

Key feelings/emotions:

Empathic highlight (formula):

It sounds like you feeling frustrated because your mother is controlling you let her be controlling

5. Young woman, 25, talking about her current boyfriend to an older confidante: "I can't quite figure him out. I still can't tell if he really cares about me, or if it's just about sex. It leaves everything up in the air. He's pleasant enough, but I'm not finding the substance that I thought was there. I do try to talk about serious things, things that interest me like what's happening in the world. The next thing I know we're back into trivia. I've been tempted to talk about this directly, but something holds me back."

Key experience(s):

feeling confused about new relationship

Key behavior(s):

expectation

Key feelings/emotions:

insecure

Empathic highlight (formula): _It sounds like you feel unsure about your new boyfriend because he does not seem to be sharing intimately with you._

6. This young man has just been abandoned by his wife after only a year of marriage: "We've been married for about a year. . . . She left a note saying that this has not been working out. Just like that. You know, I thought that things were going fairly well. Not perfect, of course. We had our ups and downs, but everyone does. Maybe I was so busy at work that there were things I didn't notice. I don't know if there's someone else. I really don't know why. . . . I must really sound stupid. I have no idea what to do to get her back. You can't make someone love you."

Key experience(s):

_____

_____

Key behavior(s):

_____

_____

Key feelings/emotions:

_____

_____

Empathic highlight (formula): _you feel hurt and rejected because your wife left suddenly and you don't know why_

EXERCISE 26: USING YOUR OWN WORDS IN EMPATHIC HIGHLIGHTS

In this exercise you are asked to do two things:

1.   First use the "you feel . . . because . . . " formula to communicate understanding to the client.
2.   Then recast your response in your own words while still identifying both core messages.

Example: A married woman, 31, is having a "solo" meeting with a marriage counselor to review how things have been going: "I can't believe it! You know when Tom and I were here last week we made a contract that said that he would be home for supper every evening and on time. Well, he came home on time every day this past week. I never dreamed that he would live up to his part of the bargain so completely! I've even begun making better meals!"

Formula. "You feel great because he really stuck to the contract!"
Non-formula. "Making good on his word has given you quite a boost!"

Now imagine yourself listening intently to each of the clients quoted below. First use the "You feel . . . because . . . " formula; then use your own words. Try to make the second response sound as natural (as much like yourself) as possible. After you use your own words, check to see if you have both a "you feel" part and a "because" part in your response.

1. Woman, 28, talking about her job with a colleague: "It's not a big thing. But this is the third time this month I've been asked to change hours with her. I said yes, of course. But it certainly seems to indicate who is more important there. Why does it always have to be me who defers to her?"

Use the formula.

You feel frustrated because your frequently having to switch hours with your co worker

Use your own words.

It sounds like your being taking advantage of at work.

2. A man who has been suffering from depression for a number of months is talking with a psychologist: "I've seen my doctor a couple of times. But now since the depression isn't getting any better, he wants to give me this drug. . . . But I don't even take aspirin if I don't have to. Somehow drugs and me don't mix. If I had pneumonia or some infection, it would be one thing. But this is all in my head. I don't want to be seen as a psycho!"

Use the formula.

You feel concerned about taking medication for depression because of what others may think

Use your own words.

your afraid to take an antidepressant because people may judge you think your crazy

3. The following trainee is talking to her instructor. The training group has just been introduced to the skill of challenging clients. They have practiced self-challenge and have just begun using challenge with one another: "For the first time in this program I'm having trouble and it's bothering me. What I've found out is that I have no problem with self-challenge. In some ways I have been doing that for years. But now I see that maybe I've been picking on myself, you know, being overly perfectionistic. And I'm finding it really hard to challenge the members of my group. Oh, I know the theory. But doing it is hard. It's almost as if they won't like me if I do it well."

Use the formula.

You feel insecure because you think your training group might be offended when you challenge them

Use your own words.

your afraid to use your challenging skills because group member might judge you.

4. A fellow trainee has been talking with you about the counselor training program. He calls you up and says he wants to talk with you right away. When you meet, he says: "You know, tomorrow we're going to start talking about our own problems. Well, the kinds of things that could stand in the way of being good helpers. . . . Well, there are things I'd rather not say in the group. . . . I even had to screw up my courage to talk with you . . . You know, a couple of issues I'd rather not talk about *could* stand in the way of my being the kind of counselor I should be. . . . But I'm just not ready. I could be dishonest."

Use the formula.

You feel inadequate because your afraid if your vulnerable w/ others they will question your counseling abilities

Use your own words.

your afraid your counseling skills will be judged if your honest about your shortcoming

5. The following person, Alfred, is being forced out of the company where he has held a senior position for almost twenty years: "It came out of nowhere. I came in two days ago and there on my desk was a notice from the venture capital group that bought us saying that I was terminated. It was as impersonal as that. . . . I kept reading it over and over again like I was in some dream. I tried to call my boss but couldn't get her. Later I found out that she had been let go, too. The notice said I get some outplacement help. What kind of word is that! It's a fancy word for dumping."

Use the formula.

you feel betrayed because your company fired you and you have been a loyal employee for 20 years.

Use your own words.

your upset and feel betrayed because your been working for that company for 20 years a loyal employee for 20 years.

6. A seventh-grade girl to her teacher, outside class: "My classmates don't like me, and right now I don't like them! Why do they have to be so mean? They make fun of me—well, they make fun of my clothes. My family can't afford what some of those dopes wear. Gee, they don't have to like me, but I wish they'd stop making fun of me."

Use the formula.

you feel hurt because some of the kids at school are making fun of you.

Use your own words.

_____

_____

EXERCISE 27: EMPATHIC HIGHLIGHTS WITH CLIENTS FACING DILEMMAS

Clients often talk about complicated issues. When this is the case, it is essential to listen even more carefully in order to pick up and respond to core messages. For example, clients often talk about conflicting values, complicated experiences, behaviors, points of view, decisions, intentions, and complex emotions. Responding with empathy means communicating an understanding of the conflict.

1.   Identify the conflict or dilemma.
2.   Use the empathic highlight formula to communicate understanding.
3.   Express the highlight in your own words.

Example: A woman, 32, talking to a counselor about adopting a child: "I'm going back and forth, back and forth. I say to myself, 'I really want a child,' but then I think about Bill [her husband] and his reluctance. He so wants our own child and is so reluctant to raise someone else's. We don't even know why we can't have children. But the fertility specialists don't offer us much hope. At times when I so want to be a mother I think I should marry someone who would be willing to adopt a child. But I love Bill and don't want to point an accusing finger at him."

> **The conflict or dilemma.** She believes that she runs the risk of alienating her husband if she insists on adopting a child, even though she strongly favors adoption.
> **Formula**. "You feel trapped between your desire to be a mother and your love for your husband."
> **You own words.** "You're caught in the middle. Adopting a child would solve one problem but perhaps create another."

1. A factory worker, 30: "Work is okay. I do make a good living, and both my family and I like the money. My wife and I are both from poor homes, and we're living much better than we did when we were growing up. But the work I do is the same thing day after day. I may not be the world's brightest person, but there's a lot more to me than I use on those machines."

The conflict.

He likes making the slary, but is not satisfied w/ his job

The formula.

You feel bored and un-satisfied because you don't think your using your full-potential on the job

Your own words.

Your torn between keeping your old job which pays well, or looking for a new more full-filling job, which will take you out 48 of your comfort zone.

2. A mental hospital patient, 54, who has spent five years in the hospital; he is talking to the members of an ongoing therapy group. Some of the members have been asking him what he's doing to get out. He says, "I don't know why you're trying to push me out of here. To tell the truth, I like it here. I'm safe and secure. So why are so many people here so damn eager to see me out? . . . Is it a crime because I feel comfortable here? (Pause, then in a more conciliatory voice.) I know you're all interested in me. I see that you care. But do I have to please you by doing something I don't want to do?"

The conflict. Pt feels comfortable and safe in hospital setting but/ and is not actively working on his TX.

The formula. you feel frustrated because the group wants you to work on being discharged from the hospital and

Your own words. your feel comfortable here, but the group is feels concerned about how long you've been here and want you to work on being discharged

3. A juvenile probation officer to a colleague: "These kids drive me up the wall. Sometimes I think I'm really stupid for doing this kind of work. They taunt me. They push me as far as they can. To some of them I'm just another 'pig.' But every time I think of quitting—and this gets me—I know I'd miss the work and even miss the kids one way or another. When I wake up in the morning, I know the day's going to be full and it's going to demand everything I've got."

The conflict. She enjoys working with teens but the attitudes of the kids/youth frustrate her.

The formula. you feel confused because you like your job but it's demanding and requires a lot of your patience and energy

Your own words. Your frustrated because the youth you work with can be challenging and disrespectful, ? put in point about liking her job.

EXERCISE 28: SHARING EMPATHIC HIGHLIGHTS IN EVERYDAY LIFE

If the communication of empathic highlights is to become a part of your natural communication style, you will have to practice it outside formal training sessions. That is, it must become part of your everyday communication style or it will tend to lack genuineness in helping situations. Practicing the sharing of highlights "out there" is a relatively simple process.

1. **Sharing empathic highlights as an improbable event.** Sharing highlights is not a normative response in everyday conversations—even among people who otherwise put into practice the value of empathy. Find this out for yourself. Observe everyday conversations. Count how many times empathy is used as a response in any given conversation.
2. **Your own use of empathy.** Next try to observe how often you yourself use empathic highlights as part of your normal style. In the beginning, don't necessarily try to increase the number of times you do so. Merely observe your usual behavior. What part does sharing highlights normally play in your interpersonal communication style?
3. **Increasing your empathic responses.** Begin to increase the number of times you share highlights. Be as natural as possible. Do not overwhelm others with this response; rather try to incorporate it gradually into your style. You will probably discover that there are quite a few opportunities for sharing highlights without being phony. Find some way of keeping track of your progress.
4. **The impact of empathy.** Observe the impact your use of highlights has on others. Don't set out to use others for the purpose of experimentation but, as you gradually increase your use of this communication skill naturally, try to see how it influences your conversations. What impact does it have on you? What impact does it have on others?
5. **Learnings.** In a forum set up by the instructor, discuss with your fellow trainees what you are learning from this experiment.

If sharing empathic highlights becomes part of your communication style "out there," then you should appear more and more natural in doing so in the training program, both in playing the role of the helper and in listening and providing feedback to your fellow trainees. On the other hand, if you use highlights only in the training sessions, it will most likely remain artificial.

Chapter 7

USING PROBES AND SUMMARIZING

Problems and opportunities are better managed when they are clear and specific. If Connie and Chuck want to deal with the poor communication they have with each other in their marriage, then the key issues relating to communication need to be spelled out in terms of specific experiences, behaviors, points of view, decisions, intentions, and feelings in specific situations. Clients need to focus on concrete rather than abstract issues. Counselors use probes to help clients "fill in the picture," turning vague statements into concrete and specific ones. Vaguely described problems lead only to vague solutions. Read Chapter Seven in *The Skilled Helper* before doing these exercises. This section has three parts: first, problem and opportunity clarity and specificity; second, probing; third, summarizing.

PROBLEM AND OPPORTUNITY CLARITY

EXERCISE 29: SPEAKING CONCRETELY ABOUT PROBLEMS

This exercise is designed to help you get a feel for both vagueness and concreteness. In this exercise you are asked to focus on your own story in terms of your own experiences, behaviors, points of view, decisions, intentions, and feelings.

1. Give two vague statements of some problem situation.
2. Turn each vague statement into a concrete and specific statement in terms of key experiences, behaviors, points of view, decisions, intentions, and feelings.

   As usual, you are asked to focus on issues—either problems or unused opportunities—that can make a difference to you in your role as helper.

Example 1: Karen writes about her need to control situations.

- **Vague statement of problem situation**: "I tend to be a bit domineering at times."
- **Concrete statement**: "I try, usually in subtle ways, to get others to do what *I* want to do. I even pride myself on this. In conversations, I take the lead. I interrupt others, jokingly and in a good-natured way, but I make my points. If a friend is talking about something serious when I'm not in the mood to hear it, I change the subject."

Example 2: Jamie talking about how the training group affects him.

- **Vague statement problem situation**: "I get anxious when I give feedback in training groups."
- **Concrete statement**: "I feel hesitant and embarrassed whenever I want to give feedback to other trainees, especially if it is in any way negative. When the time comes, my heart beats faster and my palms sweat. I feel like everyone is staring at me, though I know they're not. I feel under pressure in the group. Outside I have a live-and-let-live mentality. I'm not used to giving feedback to anyone."

In the spaces below explore two issues that are related to some problem situation or situations of your own. Try to choose issues that might affect the quality of your helping.

1. Vague statement.

_____

_____

Concrete statement.

_____

_____

_____

_____

_____

_____

2. Vague statement.

_____

_____

Concrete statement.

_____

_____

_____

_____

_____

_____

Share one or both of these with a learning partner. Get feedback on how clear your second statement is. If you do not think that your partner's statement is as clear as it might be, use probes to help him or her make the statement clearer.

EXERCISE 30: SPEAKING CONCRETELY ABOUT UNUSED OPPORTUNITIES

In this exercise, you are asked to speak about some of your unused or underused resources or opportunities. As in the exercise above, start with a vague statement, then clarify it with the kind of detail needed to serve the opportunity-development process. Try to choose unused resources or opportunities that might add value to your role as helper.

1.   Give a vague statement of some unused or underused resource or opportunity.
2.   Turn the vague statement into a concrete and specific statement in terms of key experiences, behaviors, points of view, decisions, intentions, and feelings.

Example 1: Jane, a counselor trainee, discusses her communication style.

- **Vague statement of unused opportunity**: "There is a very good communicator inside me that doesn't get out enough."
- **Concrete statement**: "This course has shown me that I am already fairly good at many of the communication skills we are learning. I listen well. I'm empathic. From what I understand about challenge, I think that I can even challenge people without turning them off. But so often I choose to be passive. It groups I'm often an onlooker rather than a participant. I found out that I can't get away with that in this group. I think I need to become much more assertive in order to be an effective learner and helper."

Example 2: Austin, in his fourth year of a doctoral program in clinical psychology, has also taken a couple of courses in organizational psychology. He has just finished his internship.

- **Vague statement of possible opportunity**: "One thing I learned during my internship rotations is that mental-health organizations are not what they could be. But I'm not sure I can do anything about it without violating my commitment to people."
- **Concrete statement**: "The organizational psychology courses I took made it clear to me that some mental-health centers are poorly-run organizations. Even with a little background I knew some things I could do to help them be more efficient and effective. In fact, the whole area of applying effective management and organizational ideas to mental-health systems appeals to me enormously. But I feel guilty about pursuing that possibility after graduation. It's like betraying the reasons for going into clinical psych in the first place—mainly to help others. I really liked dealing with individuals and groups during my internship, but I was distracted by how poorly these places are administered. A number of psychologists I met thought that administration was for the birds, you know, for lesser people."

In the spaces below, deal with two instances of unused or underused resources or opportunities. Choose opportunities that could add value to your role as helper.

1. Vague statement. *I'm a good listener but can be can easily distracted*

Concrete statement. *I have the ability to be a good listener and have learned how to focus more intently. There are times when I'm not grounded and fidget as well as loose concentration. I need to practice the new SOLER technique and make sure I am taking care of myself physically spiritually and emotionally. If I'm grounded balanced in all areas being attention will be easier*

2. Vague statement.

I can be humble, but my pride gets in the way of admitting my weaknesses

Concrete statement.

In most areas of my life I can be humble and admitt when I can't perform a task. when it comes to my physical limitations as I get older it's much harder. I need to accept that I have back problems that interfere with me being able to exercise like I used too.

Share one or both of these with your learning partner. Get feedback on how clear your second statement is. If you do not think that someone's statement is as clear as it might be, use probes to help your fellow trainee make his or her statement clearer.

PROBES

EXERCISE 31: PROBING FOR KEY ISSUES AND CLARITY

A probe is a statement or a question that invites a client to discuss an issue more fully. In the previous exercises you were asked to probe yourself. Probes are ways of getting at important details that clients do not think of or are reluctant to talk about. They can be used at any point in the helping process to clarify and explore issues, search for missing data, expand perspectives, and point toward possible client actions. An overuse of probes can lead to gathering a great deal of irrelevant information. The purpose of a probe is not data for its own sake, but information—experiences, behaviors, points of view, decisions, intentions, and feelings—that serve the process of problem management and opportunity development.

In this exercise, brief problem situations together with a bit of context will be presented. Your job is to formulate possible probes to help clients move forward in the helping process.

1. First use empathic highlights to understand the client.
2. Formulate a probe that might help the client further identify or explore key issues, experiences, behaviors, points of view, decisions, intentions, and feelings.
3. Briefly state your probe that might help the client move forward.

Example: A man, 24, complains that he is severely tempted to go on experimenting sexually with women other than his wife: "Although I have not had an extended affair, I have had a few sexual encounters and feel that some day I might have an affair. I don't blame myself or my wife, though I'm not sure I'll ever find the kind of sexual satisfaction I want with her. I've never had such strong sexual urges. I thought that once I was married, the sexual thing would be taken care of. That's certainly not the way things have turned out. I'm not sure where I'm headed."

Empathic statement: "It sounds like your sexual urges are so strong that right now they're in the driver's seat. But you're puzzled and are not quite sure what all this means and where it is going."
Possible probe: "What do you think are some of the possible consequences of the course you've taken?"
How might it help the client move forward? To help the client begin thinking through some of the implications of the decisions he's made about his sexual life.

1. Grace, 19, an unmarried, first-year college student, comes to counseling because of an unexpected and unwanted pregnancy: "Right now I realize that the father could be either of two guys. That probably says something about me right there. I'm not sure what I want to do about the baby. I haven't told my parents yet, but I think that they will be very upset, but in the end sympathetic. But I've gotten myself into this mess and I have to get myself out." *issues— unwanted pregnancy telling parents*

Empathic statement.

It sounds like your unsure about what you want to do about your pregnancy and how to tell your parents

Possible probe.

What do you consider all your choices to be decision

How might it help the client move forward?

Help her weigh pros + cons of each decision including parental support.

2. You are a counselor in a halfway house. You are dealing with Tom, 44, who has just been released from prison where he served two years for armed robbery. This was his first encounter with the law. He has been living at the halfway house for two weeks. The halfway house experience is designed to help him reintegrate himself into society. Living in the house is voluntary. The immediate problem is that Tom came in drunk a couple of nights ago. He was supposed to be out on a job-search day. Drinking is against the rules of the house. When you talk to him, he says, "I know this looks bad. But I don't think it's as bad as it looks. There was no liquor inside prison. This was my first encounter with alcohol since I got out. I guess I just went overboard. I shouldn't have done it, but it's not the end of the world. Or it shouldn't be."

Empathic statement.

It sounds like you made a poor decision and are afraid of the possible consequences

Possible probe.

What do you think you need to do next?

How might it help the client move forward?

Help client explore why he think drinking one time is not a major issue.

3. Arnie is a born-again Christian. He has begun to do a fair amount of informal preaching at his place of employment. While some of his co-workers sympathize with his views, others are turned off. Since he feels that he is being driven by a "clear vision," he becomes more and more militant. His supervisor has cautioned him a couple of times, but this has done little to change Arnie's behavior. Finally, he is given an ultimatum to talk to one of the counselor's in the Employee Assistance Program about these issues or be suspended from his job. He says to the counselor, "I have a duty to spread the word. And if I have a duty to do so, then I also have the right. I'm a good worker. In fact, I believe in hard work. So it's not like I'm taking time off for the Lord's work. Now what's wrong with that?"

Empathic statement. *It sounds like you feel concerned because your being critized for sharing important information*

Possible probe. *It sounds like you believe in your faith so much you want to carry the message but some co-workers feel offended.*

What are the possible consequences if you continue

How might it help the client move forward?

Help client identify how his continued behavior may cause him to loose his job.

4. A trainee in the counselor training program makes this statement in his small group: "I do not take criticism well. When I receive almost any kind of negative feedback, I usually smile and seem to shrug it off, but inside I begin to pout. Also, deep inside, I put the person who gave me the feedback on a 'list.' Those on the list have to pay for what they did. I find this hard to admit, even to myself. It sounds so petty. For instance, two weeks ago in the training group I received some negative feedback from Cindy. I felt angry and hurt because I thought she was my friend. Since then I've tried to see what mistakes she makes here. I've even felt a bit disappointed because I haven't been able to catch her in any kind of glaring mistake. It goes without saying that I'm embarrassed to say all this."

Empathic statement. *It sounds like you feel insecure when fellow trainees give you negative feedback and you hold a grudge afterwards*

Possible probe. *why do you think someone that critizes isnt your friend*

How might it help the client move forward?

Help client understand why she has trouble accepting criticism

5. Renata is talking with a counselor about her relationship with her mother: "I feel guilty and depressed whenever my mother calls and implies that she's lonely. I then get angry with myself for giving in to guilt

so easily. Then the whole day has a cloud over it. I get nervous and irritable and take it out on others. Or I brood. Sometimes it even interferes with my work. Then I do what I have to."

Empathic statement.

It sounds like you have a hard time when your mom feels lonely

Possible probe.

What are the possible reasons that you feel guilty

How might it help the client move forward?

Help client explore how her reaction interfere with her own serenity/happiness

EXERCISE 32: EMPATHIC HIGHLIGHTS AND TWO KINDS OF PROBES

This exercise asks you to combine several skills—the ability to share understanding through empathic highlights, to identify areas needing clarification, and to use both question-type and statement-type probes to help clients clarify their stories, points of view, decisions, and intentions together with how they feel about all of these.

1.   First reply to the client with empathy.
2.   Use a question-type probe in open-ended question form to help the client explore or clarify some issue.
3.   Convert the question-type probe into a statement-type probe.
        Indicate how the probe might help the client move forward.

Example: A law student, 25, is talking to a school counselor: "I learned yesterday that I've flunked out of school and that there's no recourse. I've seen everybody, but the door is shut tight. What a mess! I know I haven't gotten down to business the way I should. This is my first year in a large city and there are so many distractions. And school is so competitive. I have no idea how I'll face my parents. They've paid for my college education and this year of law school. I'll have to tell them that it's all down the drain."

*   **Empathic response:** "In your eyes the punishment doesn't seem to fit the crime. And it all sounds, not just desperate, but so final."
*   **Question probe:** "Which people did you see and which doors were shut?"
*   **Statement probe:** "I'm not sure who you mean by 'everybody' and it's not clear which doors are completely shut."
*   **How might the probe help the client move forward?** It might help the client identify possibilities that he has overlooked because of the emotionality surrounding the situation.

1. A high school senior to a school counselor: "My dad told me the other night that I looked relaxed. Well, that's a joke. I don't feel relaxed. There's a lull right now, because of semester break, but next semester I'm signed up for two math courses, and math really rips me up. But I need it for science since I want to go into pre-med."

Empathic response.

_____

_____

Question probe.

_____

_____

Statement probe.

_____

_____

How might it help the client move forward?

_____

_____

2. A woman, 27, talking to a counselor about a relationship that has just ended (she speaks in a rather matter-of-fact voice): "About three weeks ago I came back from visiting my parents who live in Nevada and found a letter from my friend Gary. He said that he still loves me but that I'm just not the person for him. In the letter he thanked me for all the good times we had together these last three years. He asked me not to try to contact him because this would only make it more difficult for both of us. End of story. I guess I've let my world collapse. People at work have begun complaining about me. And I've been like a zombie most of the time."

Empathic response.

_____

_____

Question probe.

_____

_____

Statement probe.

_____

_____

How might it help the client move forward?

_____

_____

3. A married man, 25, talking to a counselor about trouble with his mother-in-law: "The way I see it, she is really trying to destroy our marriage. She's so conniving. And she's very clever. It's hard to catch her in what she's doing. You know, it's rather subtle. Well, I've had it! If she's trying to destroy our marriage, she's getting pretty close to achieving her goal."

Empathic response.

_____

_____

Question probe.

_____

_____

Statement probe.

_____

_____

How might it help the client move forward?

_____

_____

4. A woman, 31, talking to an older woman friend: "I just can't stand my job any more! My boss is so unreasonable. He makes all sorts of silly demands on me. The other women in the office are so stuffy, you can't even talk to them. The men are either very blah or after you all the time, you know, on the make. The pay is good, but I don't think it makes up for all the rest. It's been going on like this for almost two years."

Empathic response.

_____

_____

Question probe.

_____

_____

Statement probe.

_____

_____

How might it help the client move forward?

_____

_____

5. A man, 45, who has lost his wife and home in a tornado, has been talking about his loss to a social worker: "This happened to a friend of mine in Kansas about ten years ago. He never recovered from it. His life just disintegrated and nobody could do anything about it. . . . It was like the end of the world for

him. You never think it's going to happen to you. I know I belong here. But this kind of thing makes me think that I don't."

Empathic response.

_____

_____

Question probe.

_____

_____

Statement probe.

_____

_____

How might it help the client move forward?

_____

_____

EXERCISE 33: PROBING FOR ACTION

Remember that each step of the helping process should be, all things considered, some kind of stimulus for client action, the "little" actions, as it were, that precede formal action based on a plan. In this exercise you are asked to use both empathic highlights and probes to help clients consider actions they might take to move forward. Probes are very useful in helping clients develop a solution-focused approach in the helping process.

Example. A divorced woman, 44, talking to a counselor about her drinking. This is the second session. She has spent a lot of time telling her story. She has already admitted that she engages in evasions and sometimes simply lies about her problems. She says, "Actually, it's a relief to tell someone the truth. I don't have to give you any excuses or make the story sound right. I drink because I like to drink; I'm just crazy about the stuff, that's all. But I'm under no delusions that telling you is going to solve anything. When I get out of here, I know I'm going straight to a bar and drink. Some new bar, new faces, some place they don't know me."

- **Empathic response**. "So, for you, drinking is a way of life. No use making excuses. That's the way it is."
- **Probe for action**. "I wonder what it would be like to do something else rather than go to a bar after we finish here. Perhaps something a bit more creative. What might that be?"

1. A man, 57, talking to a counselor about a family problem: "My younger brother, he's 53, has always been a kind of bum. He's always poaching off the rest of the family. Last week my unmarried sister told me that she'd given him some money for a 'business deal.' Business deal, my foot! I'd like to get hold of him and give him a good kick! Oh, he's not a vicious guy. Just weak. He's never been able to get a fix on life. But he's got the whole family in turmoil now, and we can't keep going through hell for him."

Empathic response.

_____

_____

Probe for action.

_____

_____

2. Graduate Student, 25, to an instructor whom he trusts "I have two term papers due tomorrow. I'm giving a report in class this afternoon. My wife is down with the flu. And now I find out that a special committee wants to talk with me about my progress in the program. I've been working a couple of part-time jobs—I've got to just to keep things going. I know I'm behind, but I still think I can handle things. I hope they're not going to give me ultimatums."

Empathic response.

_____

_____

Probe for action.

_____

_____

3. A man, 49, talking to a rehabilitation counselor after an operation that has left him with one lung: "I'll never be as active as I used to be. But at least I'm beginning to see that life is still worth living. I have to take a long look at the possibilities, no matter how much they've narrowed. I can't explain it, but there's something good stirring in me."

Empathic response.

_____

_____

Probe for action.

_____

_____

4. Mark and Lisa, a married couple, both 33, after years of attempting to have children finally adopted a baby girl, Andrea. Their relationship, which up to then seemed quite good, has begun to disintegrate. Mark has made some cracks about "the stranger in the house." Andrea has proved to be a somewhat difficult baby. Lisa feels exhausted and blames Mark for not helping her. They are both thinking about divorce now, but feel very guilty because of the child. Both of them say, "If only we had never adopted Andrea."

Empathic response.

_____

_____

Probe for action.

_____

_____

EXERCISE 34: USING PROBES TO HELP OTHERS COMMUNICATE BETTER

This is an exercise that can be done both in the training group and in your everyday life. You most likely run into people who are not very good at dialogue. You can use probes to help them do better.

1.   Find a person who is not a good communicator, someone not good at dialogue.
2.   Use both highlights and probes to help this person engage in dialogue, that is, engage in turn taking, connecting with what you have to say, staying open to mutual influence, and, in the end, co-creating the conversational outcome _with_ you.
3.   After you have engaged in a couple of these conversations, discuss your successes and your failures with a learning partner. What do you have to do to become better at helping less-gifted people engage in dialogue? Ultimately, this means helping clients engage in dialogue as they tell their stories, reveal their points of view, discuss their decisions, and reveal their intentions.

Example. Trudy, a counselor trainee, is talking with Nathan, a friend of hers in the neighborhood. Nathan has been talking almost nonstop about why he thinks that young people ignore politics and don't vote. Trudy uses the skills she is learning to try to help Nathan have a dialogue with her.

> **Nathan** (continuing at a fast pace): Take the last presidential election.
> **Trudy** (interrupting): Nathan, your main point seems to be that politicians, because of the crazy things they do, don't deserve the attention of young people. To me that makes young people sound a bit too noble. I think that we—I include you and me among the young—take our country and what it offers too much for granted.
> **Nathan**: What do you mean by that?
> **Trudy**: Well, we're a bit naive and selfish. As long as we get what we want, we just leave it to the politicians to run the country. We're just not interested.
> **Nathan**: Who cares anyway. In the last presidential election the country joined us and said no to both of them. . . .
> **Trudy**: What about my point—that we can't just blame the politicians? We have some responsibility for this drama. It's easy for us to take potshots at politicians.
> **Nathan**: Well, we're not kids. We're actually grown ups. But politicians don't listen to us anyway.
> **Trudy**: We're grown up physically, but are we grown up politically?
> **Nathan**: Some young people are. Look at Carlos. He makes more sense to me about politics than the politicians.
> **Trudy**: Right. But Carlos seems to be the exception rather than the norm.
> **Nathan**: Well, if more of us were like Carlos, it would make a huge difference.
> **Trudy**: What kind of difference?
> **Nathan**: We could swing elections. Certainly the last presidential election.
> **Trudy**: So we have the ability to be more politically mature than we are, but not many of us make it a priority.
> **Nathan**: Hey, we've got a lot of things to do. Look around for a partner or mate. Get degrees. Consider careers. Not all of that is selfish, you know. It's life.
> **Trudy**: So there are serious reasons for us not to be as politically mature as we might be. There are real life tasks. And they're not just self-distractions or signs of immaturity.

And on they go. Discuss Trudy's approach here with a learning partner. What did she do right? What did she do wrong? What could she have done to improve the quality of the dialogue?

SUMMARIZING

At a number of points throughout the helping process it is useful for helpers to summarize or to have clients summarize the principal points of their interactions. This places clients under pressure to focus and move on. Since summarizing can place pressure on clients to move forward, it is sometimes a form of challenge.

EXERCISE 35: SUMMARIZING AS A WAY OF PROVIDING FOCUS

This exercise assumes that trainees have been using the skills of sharing highlights and probing in order to help one another tell their stories, share their points of view, discuss their decisions, and spell out their intentions.

1.   The total training group is divided into subgroups of three.
2.   There are three roles in each subgroup: helper, client, and observer.
3.   The helper spends about eight minutes counseling the client. The client should continue to explore one of the problem areas he or she has chosen to deal with in the training group.
4.   At the end of four minutes, the helper summarizes the principal points of the interaction. Helpers should try to make the summary both accurate and concise. The helper can draw on past interactions if he or she is counseling a fellow trainee whom he or she has counseled before. At the end of eight minutes or so, the helper should engage in a second summary.
5.   At the end of each summary, the helper should ask the client to draw some sort of implication or conclusion from the summary that helps the client move forward.
6.   Then both observer and client give the helper feedback as to the accuracy and the helpfulness of the summaries provided. The summaries are helpful if they help move the client toward problem clarification, goal setting, and action.
7.   This process is repeated until each person in the subgroup has had an opportunity to play each role.

Example 1: It would be too cumbersome to print eight minutes of dialogue here, but consider this brief outline of a case. A young man, 22, has been talking about some developmental issues. One of his concerns is that he sees himself as relating to women poorly. One side of his face is scarred from a fire that occurred two years previous to the counseling session. He has made some previous remarks about the difficulties that he has relating to women. After four minutes of interaction, the helper summarizes:

> **HELPER**: Dave, let me see if I have the main points you've been making. First, because of the scars, you think you turn women off before you even get to talk with them. The second point, and I have to make sure that this is what you are saying, is that your initial approach to women is cautious, or cynical, or maybe even subtly hostile since you've come to expect rejection somewhat automatically.
> **DAVE**: Yeah, but now that you've put it all together, I am not so sure that the hostility is all that subtle.
> **HELPER**: So at least at times there is some open hostility. . . . You also said that the women you meet are cautious with you. Some might see you as a bit nasty, I believe that was your word. Others steer clear of you because they see you as difficult to talk to and get to know. Finally,

what closes the circle is that you interpret their caution as being turned off by your physical appearance.

DAVE: I don't like to hear it that way, but that's what I've been saying.

HELPER: If these points are fairly accurate, I wonder what implication you might see in them.

DAVE: Well, one possibility is that *I'm* the one that rejects me because of my face. If that's the case, then nothing's going to get better until I come to grips with myself.

Note that the client draws an implication from the summary ("I am the primary one who rejects me") and moves on to some minimal declaration of intent ("I need to change this").

Example 2: Artemis, a 37-year-old woman, has been talking about her lack of assertive behavior in a paraprofessional helper training group. She sees clearly that his lack of assertion stands in the way of being an effective helper. She and her helper explore this theme for a few minutes and then the helper invites her to give a summary of the main points she is making:

> **HELPER:** Artemis, it might help if you would pull together the main points you've been making. It might help us to see the way forward.
>
> **ARTEMIS:** Okay, let's see if I can pull this together. I'm convinced that the ability to move into the life of the client without apologizing for it is essential. However, this simply has not been part of my normal interpersonal style. I don't intrude into the lives of others. If anything, I'm too hesitant to place demands on anyone. When I take the role of helper in training sessions, I feel awkward even sharing basic highlights and even more awkward using probes. As a result, I let my clients ramble and their problems remain unfocused. In social settings I'm quite passive, except now I'm much more aware of it.
>
> **HELPER:** What implication might you draw from it?
>
> **TRAINEE:** When I put it all together like that, my immediate reaction is to say that I shouldn't try to be a counselor. But I think I would be selling myself short. No matter what career I choose to follow, I need to become more assertive. I have to learn how to take risks with people.

After each trainee gives his or her summary and elicits some reaction from the client, the feedback from the client and the observer should center on the accuracy and the usefulness of the summary, not on a further exploration of the client's problem. Recall that feedback is most effective when it is clear, concise, behavioral, and non-punitive.

EXERCISE 36: HELPING OTHERS ENGAGE IN DIALOGUE

The skills of probing and summarizing can be used to assist hesitant clients to become partners in the helping dialogue. Summarizing lets the other person know you are listening to them in a thoughtful manner. By itself this encourages clients to open up. Probes both direct the client's attention to important matters and indicate the interest of the helper. Combined summarizing and probing can help a client, even a hesitant client, become a fuller partner in the dialogue.

Example: Sophie, 63, has been divorced from her husband for many years. Their marriage was difficult and her often blamed her cruelly for their inability to have children. She always assumed that she was at fault and for years absorbed his abuse feeling she had no recourse. Recently, she learned from the family doctor that it was he that was sterile not she. He knew the truth but misled her. "I didn't know what to say. For years I thought I was to blame and I felt so bad about it I gave up any right to resist his demands and his abuse. I suppose he had his reasons, but I can't tell you how many nights I cried. I don't even know where to begin; this seems so . . ."

Summary: "It's terribly unsettling to learn that something which shaped a lot of what went on between you and your husband was, well, based on a lie. You spent many years believing your inability to have children brought unhappiness to your husband and now you find that it wasn't that way at all."
Probe: "Given what you know now, how does this affect the way you see yourself?"

1. Identify people you know who are not good communicators. That is, people who are not effective at getting their story across to others. Limit this to one or two individuals you see on a regular basis.
2. Describe to your learning partner who these individuals are and how you might improve their communication if you were to employ the skills of summarizing and probing.
3. In the next week find an opportunity to engage them in conversation and make your best effort to listen, summarize, and probe.
4. Report back to your learning partner on the experience. What worked? What didn't work? What impact did you have on this other person?

PART THREE

STAGE I OF THE HELPING MODEL
AND ADVANCED COMMUNICATION SKILLS

The basic communication skills reviewed in Part Two are critical tools.

With them, you can help clients engage in all the stages and steps of the

helping model. But those communication skills are not the helping

process itself. Part Three is a detailed exposition and illustration of

Stage I of the helping model, together with its three steps. Step I-A,

helping clients tell their stories is discussed in Chapter 8. Chapter 9

deals with reluctant and resistant clients. Step I-B focuses on advanced

communication skills— those related to helping clients challenge

themselves. These skills are discussed and illustrated in Chapters 10, 11,

and 12. Chapter 13, Step I-C, introduces the concept of leverage—

helping clients choose the right issues to work on.

Clients come to helpers because they need help in managing their lives more effectively. Stage I illustrates three ways in which counselors can help clients understand themselves, their problem situations, and their unused opportunities.

> **Step I-A: Stories**. Help clients tell their stories, share their points of view, explore their decisions, and clarify what they intend to do together with the emotions involved.
> **Step I-B: Blind Spots**. Help clients identify and move beyond blind spots to new perspectives that will help them manage their problem situations and opportunities more effectively.
> **Step I-C: Leverage**. Help clients choose to work on issues that will most likely make a difference in their lives.

As indicated in *The Skilled Helper*, these principles are interrelated and not restricted to Stage I for the following reasons. First of all, clients don't tell their full stories at the beginning of the helping process. Often the story "leaks out" over time. Second, blind spots can appear at any stage or step of the helping process. Blind spots affect choosing goals, setting strategies, and implementing programs. Third, leverage deals with the "economics" of helping. Choosing the right problem or opportunity to work on is just the beginning. Choosing the right goals, strategies, and resources for implementation is also part of the picture.

Chapter 8

STEP I-A:
HELPING CLIENTS TELL THEIR STORIES:
"WHAT ARE MY CONCERNS?"

Step I-A deals with helping clients tell their stories, that is, discuss the problem situations, anxieties, and concerns that bring them (or get them sent) to the helper in the first place and identify and discuss unused opportunities. Read Chapter Eight in the text before doing the exercises in this section.

EXERCISE 37: HELPING ONE ANOTHER TELL YOUR STORIES

In the exercises you have done to this point you have already been helping clients tell their stories through the use of empathic highlights, probes, and summaries. It is time now to switch from writing to dialogue.

1. Form groups of three.
2. In each group there are three roles: client, helper, and observer.
3. Before the dialogue between client and helper begins, the trainee in the client role gives a brief summary of the problem situation or unused opportunity he or she is going to explore.

Example: Enrico, the trainee playing the role of client, gives the following summary: "I'm going to discuss a problem I'm having with my parents. I'm an only child and they simply won't let go. Don't get me wrong. They're very nice about it. But I'm 21. We're not a wealthy family. So, for the time being, I have to live at home. Also I believe that becoming my own person is directly related to becoming an effective helper."

4. The helper begins by saying: "Against that background, Enrico, what would you like to focus on today."
5. Then client and helper engage in a helping dialogue. The helper, using whatever combination of empathic highlights, probes, and summaries he or she sees necessary, helps the client tell his or her story clearly and in useful detail.
6. After about six minutes the observer, in timekeeper mode, stops the dialogue.
7. The client gives the helper feedback on how effectively he or she has aided the story telling process. Since giving feedback to your fellow learners is an important part of the training program, review the section in the Introduction on guidelines for giving effective feedback.
8. Then the observer shares his or her observations.
9. The process continues until each member has had the opportunity to play each role.

EXERCISE 38: TALKING PRODUCTIVELY ABOUT THE PAST

Clients will of necessity talk about what has happened to them in the past. In some cases the client uses the past to add depth to their story or to supply needed context. However, at other times talking about the past without an effort to make a link to what is presently happening in the client's life can inhibit the helping process.

1. In this exercise listen to your partner discuss a past event in his or her life.
2. Use highlights, probes, and summaries to help his or her partner talk about the past. Invite the client to draw from the past event some implications for the present.
3. When you have finished discuss the dialogue with your partner. To what extent did the probes and/or summary help make the connection to the present? What else could you have done to help your partner see the implications of the past for the present?

Example: Picture yourself talking with a man, 33, who has just started dating someone seriously. He is reflecting on a past relationship that didn't work out. In your dialogue with him you learn the following: "I guess I didn't realize it then, but she was a really great woman, one of the best I will ever know. We fell for each other and everything was great. Then we had this fight. I took a very hard stand and made it impossible for her. God, I was stubborn. Angry, too. Then I began seeing other women. Of course, she found out. I might have salvaged it even then if I had gone to her and said what I really felt. But, I just did what I always do—called up some buddies and went drinking." Of course in the real conversation you would have been sharing highlights along the way. Here are some things you might have eventually said:

> **Response A**: "I wonder how you might use what you've learned from this past experience to prevent something similar happening in your new relationship."
> **Response B**: "It sounds like you learned—somewhat painfully—something very useful from that relationship. What did you learn?"
> **Response C**: "If you had that relationship to do over again, what would you do differently?"

EXERCISE 39: LINKING STORYTELLING TO ACTION

As noted in the text, clients need to engage in "little" actions throughout the helping process that get them moving in the right direction even before formal goals and action strategies are established. These little actions are signs of clients' commitment to the process of constructive change. In this exercise you are asked to put yourself in the place of the client and come up with some possible actions that the client might take to move forward in some way. It is not that you are going to tell the client what to do. But helping clients adopt an action orientation to their problem situations is central to helping.

1. Listen to the story.
2. If you were the client, what are two common-sense actions you might consider taking at this stage in order to move forward. Even if "big" actions seem premature, what "little" actions might you take?
3. Indicate the reason for each possible action.
4. Get feedback from a partner on the actions and the reasoning behind them.

Example. A seventh-grade boy talking to a teacher he trusts (all this is said in a halting voice and he does not look at the teacher): "Something happened yesterday that's bothering me a lot. I was looking out the window after school. It was late. I saw two of the guys, the bullies, beating up on one of my best friends. I was afraid to go down there. . . . A coward. . . . I didn't tell anyone, I didn't do anything."

One possible action: First, the boy might talk the incident through with his friend and perhaps apologize for letting him down
Reason: to reestablish relationship with his friend and to show basic decency.
Another possible action: Second, the boy might figure out how he wants to handle similar situations in the future.
Reason: to learn something useful from this painful experience.

1. A young woman talking to a counselor in a center for battered women. Other counselors have seen her on two different occasions: "This is the third time he's beaten me up. I didn't come before because I still can't believe it! We're married only a year. After we got married, he began ordering me around in ways he never did before we got married. He'd get furious if I questioned him. Then he began shoving me if I didn't do what he wanted fast or right. And I just let him do it! I just let him do it! (She breaks down and sobs.) And now three beatings in less than two months. Oh God, what's happened?"

Two possible client actions and a reason for each.

_____

_____

_____

_____

_____

2. A man, 70, arrested for stealing funds from the company where he has worked for 25 years, talking to an assigned social worker. This is their second session: "To tell you the truth, it's probably a good thing I've been caught. I've been stealing on and off for the last five or six years. It's been a game. It soaked up my energies, my attention, distracted me from thinking about getting old. Now I'm saying to myself: 'You old fool, what're you running from?' You're probably thinking: 'It's about time, old guy.'"

Two possible client actions and a reason for each.

_____

_____

_____

_____

_____

3. A girl, 15, talking to a psychologist at a time when her parents are involved in a divorce case: "I still want to do something to help, but I can't. I just can't! They won't let me. When they would fight and get real mean and were screaming at each other, I'd run and try to get in between them. One or the other would push me away. They wouldn't pay any attention to me at all. They're still pushing me away. They

don't care how I think or feel or what happens to me! My mother tells me that kids should stay out of things like this."

Two possible client actions and a reason for each.

_____

_____

_____

_____

_____

4. A young man, 24, is talking to his mentor, a senior partner in the company. A department head has offered him a managerial job in her area. In this job he will be leading an important project. He says, "I didn't think this kind of opportunity would come so soon. But I've decided to say yes. I'm a bit worried, sure, but I figure I was asked because they think I can do the job. You know, just three years ago I was studying this stuff in school and now I'm going to have a budget, people reporting to me, and a chance to make a mark. I have a pretty clear idea about what I'm going to do when I get started next week. Most of all I want to get off on the right foot with the people."

Two possible client actions and a reason for each.

_____

_____

_____

_____

_____

5. A woman, 34, a local news reporter has completed a series on an AIDS hospice. She talks with the hospice volunteer director. "In my job I get to see a lot and there is a tendency to grow cynical. But, what you do here and how you do it is staying with me like few stories I've done. I'd like to give some time as a volunteer, but I don't want any public recognition. I'd just rather do the work as quietly as possible. I don't want to seem to be doing this for the wrong reasons. Personally, I need to be grounded in people's lives in a real way, but my job makes that difficult. Somehow I think this is the kind of work where I can contribute something as a person and not as a reporter."

Two possible client actions and a reason for each.

_____

_____

_____

_____

_____

_____

EXERCISE 40: TAKING ACTION IN YOUR OWN LIFE

This exercise challenges you to experience the effect of taking action. In the previous exercise you imagined what actions a client might take. In this exercise you are invited to make it real by taking an action, large or small, in your own life. You and your partner can engage in this exercise at the same time.

1. Identify a problem or opportunity which is ripe for action of some sort.
2. Discuss with your partner what "little actions you could take—actions that would make things better or move the problem-managing process forward. Talk them through with your partner.
3. Take action "out there" in your life.
4. Come back and discuss with your partner the effect of taking action.

Chapter 9

MANAGING RELUCTANCE AND RESISTANCE

In order to recognize reluctance and resistance in clients, it helps to first be able to recognize the same tendencies within yourself and learn how to manage them in your own life. Read Chapter Nine in *The Skilled Helper* before doing the following exercises.

EXERCISE 41: MY OWN RELUCTANCE

Since living more effectively requires work, is sometimes painful, and entails risk, all of us express reluctance from time to time. This exercise targets your own reluctance to become more than you are.

1.   Review the problems, unused or underused opportunities and resources, and developmental challenges in your life.
2   Identify three areas or ways in which you have been reluctant to grow or change. Spell them out in some detail. Note the difference between reluctance and resistance.

Example. Michaela, 19, a first-year student in clinical psychology, has this to say about developing a sense of responsibility: "The more I learn about the world, the more I see how comparatively easy life has been for me. Everything has been given to me. I'm a product of the American dream. I had an easy time in high school, but the first months of graduate school have been jarring. I'm coming to realize that no one is going to give me anything. As I look down the road I see that grad school more accurately reflects the real world. Without getting into useless guilt trips, I realize I'm spoiled. I also realize that old habits do not die easily. I still expect things to be handed to me on a platter. I'm reluctant to face up to the realities of imposed schedules, competition, demanding work, and budgets. It is very difficult for me to accept that some of my learning is going to come from making mistakes, being challenged, and having to do things over."

First area of reluctance.

_____

_____

_____

Second area of reluctance.

_____

_____

_____

Third area of reluctance.

_____

_____

_____

3.　　Choose one of the areas and indicate what you might do in order to overcome your reluctance.

_____

_____

_____

_____

_____

_____

4.　　As you share these instances of personal reluctance with a learning partner or the entire training group, see if you can identify any underlying themes that run through all three areas.

_____

_____

_____

EXERCISE 42: MY OWN EXPERIENCE OF RESISTANCE

Recall that resistance is a reaction to someone's trying to force you to change when you don't want to. You may even think that the other person has a point, but the fact that he or she is telling you what to do causes you to react rather than respond.

1.　　Recall two instances when you resisted growth because you felt you were being forced into it.
2.　　Indicate how you could have transformed that negative experience into a positive one for you.

Example. Manfred, a 24-year-old graduate student in social work, recalls being badgered by a former girl friend: "Elise was out to reform me. It's not that I was not in need of reform during my college years. But the way she went about it turned me off. For instance, I drank too much. But she lectured me in private. Then she embarrassed me in front of my friends—her way of getting my attention. Before going out to major events like the homecoming dance, she laid down rules and made me promise to keep them. Otherwise she wouldn't go out with me. Things like that."

When asked how that negative experience could have been turned into a positive one, he had this to say: "The substance of what she was saying actually made sense. But I would have done better if she had backed off a bit, quite a bit at times. For instance, instead of laying down rules to prevent disasters, she could have quietly left the scene if they did occur. If she had said that she was no longer having fun and was going home and then just did it a couple of times, I think that would have been a real wake up call for me. I could have said all of this to her, but I never did. I just reacted."

The first instance of resistance in your own life.

_____

_____

_____

Indicate how that negative experience could have become a positive one for you.

_____

_____

_____

_____

_____

The second instance in your on life.

_____

_____

_____

Indicate how that negative experience could have become a positive one for you.

_____

_____

_____

_____

_____

EXERCISE 43: YOUR OWN WAYS OF CAUSING RESISTANCE

Resistance in clients is prompted by real or perceived coercion on your part. At times you may not think that you are putting any undue pressure on a client, but he or she might feel that you are. That is, you don't intend to be coercive but nevertheless you come across that way. For example, let's assume that you are very open about yourself. That's the way you grew up. However, when you place a demand on others for higher levels of self-disclosure than they are ready for, they may clam up. Therefore, there is much to be gained from knowing what aspects of your own style might come across to others as subtly, or not so subtly, coercive.

1.   Jot down some aspects of your style that might be seen as coercive by others.
2.   Discuss with your partner what aspects of your helping style could be a problem for certain clients.
3.   Finally, talk about ways of changing or softening your style.

Coercive aspects of your style.

_____

_____

_____

_____

_____

Changes that would help.

_____

_____

_____

_____

_____

EXERCISE 44: DEALING WITH RESISTING CLIENTS

You need to be able to identify the signs that you are stirring up resistance in clients. Some clients start pushing back. Others begin to drag their feet. Then, once you discover that you are putting clients off, you need to recover.

Example: Marisol, 42, is an experienced counselor working in a health-care facility. Recently, she transferred within the hospital from the emergency room—where she handled individuals and families in crisis—to the oncology unit. She talks about how her emergency-room style has caused problems in the oncology unit: "Down in the ER we had to work fast and I got good at making things happen. Now I'm in oncology and I'm getting on people's nerves. I feel like I'm out of synch. I want to help both staff and patients make decisions but our timetables are all different. I'm moving fast—the way I did in the emergency room—I want people to move at my pace. I'm scaring some people. Others are grateful for my assertiveness and candor. I need to tailor my style to the unit, to the different situations, and to different people. I have to pay close attention to what people need and then pace myself."

1. Imagine you are working with a client and it appears that they are resisting. Describe the situation.

_____

_____

_____

_____

_____

2. What are some of the things you could do to recover?

_____

_____

_____

_____

_____

Discuss your findings with a learning partner.

Chapter 10

STEP I-B: HELPING CLIENTS CHALLENGE THEMSELVES

Chapter Ten in *The Skilled Helper* discusses the role of challenging in the helping process. Helping clients challenge themselves both to participate actively in the helping process and to act on what they learn adds great value. However, while challenge without understanding is abrasive, understanding without challenge is anemic. Before doing the following exercise, read Chapter Ten.

We begin with an exercise in self-challenge. Many of the clients you see will be struggling with developmental concerns like those you have struggled with or are currently working on. Like you, they will have both strengths and "soft spots" in coping with the developmental tasks of life.

EXERCISE 45: REVIEWING SOME BASIC DEVELOPMENTAL TASKS

In this exercise you are asked to consider your experience with ten major developmental tasks of life. First reflect on your experience in these developmental areas and then apply what you have learned to your role as a helper of others. Use extra paper as needed.

1. **COMPETENCE. What do I do well?** To what extent do I see myself as a person who is capable of getting things done? How effective am I in gathering the resources needed to accomplish the goals I set for myself? In what areas of life do I excel? In what areas of life would I want to be more competent than I am?

Strengths	Soft Spots
_____	_____
_____	_____
_____	_____
_____	_____
_____	_____

2. **AUTONOMY: How well do I make it on my own**? How effectively do I avoid being either overly dependent or too independent? To what degree am I reasonably interdependent in my work and social life? When I need help, how easy do I find it to ask for it? In what social settings do I find myself most dependent? counterdependent? independent? interdependent?

Strengths	Soft Spots
_____	_____
_____	_____
_____	_____
_____	_____

3. **VALUES: What do I believe in and prize?** What are my principal values? To what degree do I pursue reasonable development in my value system? How well do I put my values into practice? What value conflicts do I face? In what social settings do I pursue the values that are most important to me?

Strengths	Soft Spots
_____	_____
_____	_____
_____	_____
_____	_____
_____	_____

4. **IDENTITY: Who am I in this world?** To what extent do I have a sense of who I am and the direction I'm going in life? What kind of congruence exists between the way I see myself and the way others see me? What kind of "center" gives meaning to my life? In what social settings do I have my best feelings for who I am? In what social settings do I tend to lose a sense of who I am? In what ways am I confused or dissatisfied with who I am?

Strengths	Soft Spots
_____	_____
_____	_____
_____	_____
_____	_____

5. **INTIMACY. What are my closer relationships like?** What kinds of closeness do I have with others? To what extent are there degrees of closeness in my life--from acquaintances to friends to intimates? What is my life in my peer group like? How well do I get along with others? What concerns do I have about my interpersonal life?

Strengths	Soft Spots
_____	_____
_____	_____
_____	_____
_____	_____

6. **SEXUALITY. Who am I as a sexual person?**   To what degree am I satisfied with my sexual identity, my sexual preferences, and my sexual behavior? How do I handle my sexual needs and wants? What social settings influence the ways I act sexually?

Strengths	Soft Spots
_____	_____
_____	_____
_____	_____
_____	_____
_____	_____

7. **LOVE, MARRIAGE, FAMILY. What are my deeper commitments like?** What is my marriage or deeper relationship like? How do I relate to family and relatives, however defined? How do I feel about the quality of my family or friendship-circle life? If not married, in what ways do I look forward to marriage or to some deeper stable relationship? What misgivings do I have about such a relationship?

Strengths	Soft Spots
_____	_____
_____	_____
_____	_____
_____	_____

8. **CAREER. What is the place of work in my life?** How do I feel about the way I am preparing myself for a career or the career I am currently pursuing? What do I get out of work? What am I like in the workplace? How does the workplace affect me? What impact do I have there?

Strengths	Soft Spots
_____	_____
_____	_____
_____	_____
_____	_____

9. **INVESTMENT IN THE WIDER COMMUNITY**. How big is my world? How do I invest myself in the world outside of friends, work, and the family? What is my neighborhood or community like? What kind of concerns do I have about community, civic, political, social issues? In what ways am I optimistic about the world? In what ways am I cynical?

Strengths	Soft Spots
_____	_____
_____	_____
_____	_____
_____	_____
_____	_____

10. **LEISURE. What do I do with my free time?** Do I feel that I have sufficient free time? How do I use my leisure? What do I get out of it? In what social settings do I spend my free time?

Strengths	Soft Spots
_____	_____
_____	_____
_____	_____
_____	_____
_____	_____

Many of the clients you see will be struggling with similar developmental concerns. Like you, they will have both strengths and "soft spots" in these areas.

1. Pick two of the ten developmental areas. Try to pick areas that will make a difference in your work as a counselor.
2. Indicate one or two new perspectives generated because of this exercise.
3. Outline the kinds of actions these new perspectives might drive.

Example. Mickey, a 20-year-old undergraduate majoring in psychology, had this to say:

*   Developmental areas: "I chose identity and investment in the wider community."
*   New perspective: "There is a huge difference between identity and what I now see as individualism. I think I've been on the selfish, me-first, what's-in-it-for-me track. I think little about the wider community and belonging to it."
*   Action implications: "First of all, I have to clear up in my mind the difference between knowing who I am, what I stand for, and where I'm going and my me-first attitudes. Second, I'd like to begin to take a critical look at the messages the society of which I am a part is screaming at me through music, movies, television, peer interaction, the internet--the whole package. A lot of the messages would keep me an adolescent and I don't want to remain an adolescent."

Now explore two areas for yourself.

First developmental area.

_____

New perspectives.

_____

_____

_____

Link these new perspectives to action.

_____

_____

_____

_____

_____

Second developmental area.

_____

New perspectives.

_____

_____

_____

Link these new perspectives to action.

_____

_____

_____

Share your learnings with a partner from your training group.

EXERCISE 46: IDENTIFYING DIFFERENT AREAS NEEDING CHALLENGE

Challenge focuses on dysfunctional mindsets and behavior, whether internal or external. There are various ways in which clients need to be helped challenge themselves in order to develop the kinds of new perspectives that lead to constructive problem-managing and opportunity-developing action. In this exercise you are asked to do the following:

1.   Read the case and determine in what area and in what way this client might benefit from challenge.
2.   Sketch out briefly what some helpful new perspectives might look like.
3.   Indicate what kind of constructive actions might flow from the new perspective.
4.   Share your observations with a learning partner. Note the similarities and the differences among your suggestions.
5.   Discuss how you might proceed with each client. Discuss ways of helping these clients challenge themselves.

Example: Hannah, a new counselor working in a hospital setting for a number of months, has begun to confide in one of her colleagues who has worked in health-care facilities for about seven years. Hannah has intimated, however tentatively, that she feels that some of the medical doctors tend to dismiss her because she is new, because she is a counselor and not a medical practitioner, and because she is a woman. One doctor has been particularly nasty, even though he masks his abusiveness under his brand of "humor." One day her colleague says to her, "Don't take that stuff. Tell 'em what you think!" Hannah replies: "That will only make them worse. I think it more important to stand up for my profession by just being who I am and helping patients as much as possible. Their meanness is a punishment in itself. Anyway, it's a game that doctors play with everyone else in the hospital. That kind of thing has been going on forever. It's not such a big deal anyway."

> **What mindsets and behavior might be challenged here?** Under the guise of "reasonableness," Hannah could be into excuse making. "This is the way it is and the way it will always be" seems to be her stance. Furthermore, Hannah says nothing about the possible impact of hospital politics on patients.
> **What would some helpful new perspectives look like**? Hannah might be helped to see that accepting the status quo goes counter to the kind of professionalism that should characterize the hospital and that she prides herself in. She may also be helped to see that the hospital's caste system might interfere with team approaches to helping patients.
> **What actions might such new perspectives lead to?** Hannah might bring the core issues before the hospital senate, she might quietly lobby some of her colleagues to see what kind of consensus there is, she might talk to the especially abusive doctor privately and let him know how she feels, she might confide in one of the doctors who seems to be free of these prejudices in order to see what course of action she might follow, and so forth.

It goes without saying that the best way of doing this exercise is through actual interaction with these clients. At this stage, however, you are being asked to do some educated guessing.

1. Dan is in a coaching/counseling session with his supervisor, June. A project is behind schedule and a key customer is complaining. Dan was chosen to be the project leader because he had manifested "leadership" qualities in his work. He says something like this: "Well, this is one of the first team-focused projects we've had. We're not used to working like this. The key people in design and engineering have been slow to come to the table. They keep telling me that they can't let other projects slip. I know the customer wants the prototype by the end of the month, but that's his schedule, not necessarily reality. He's so pushy. Lately, I find it a real chore just talking to him."

What mindsets and related behavior might be challenged here?

_____

_____

_____

What would some helpful new perspectives look like?

_____

_____

_____

What actions might such a perspective lead to?

_____

_____

_____

2. Claudette and Jim have tried unsuccessfully to have children. Doctors have examined both of them and now offer little hope even with new medical techniques. They have both been seeing a counselor, but in this session Claudette, who is much less reconciled to their fate than Jim is, is seeing the counselor alone. She says, "It's just not fair. Doctors talk about infertility as a clinical phenomenon, not a human reality. Jim and I love one another. We want to have children. We'd make very good parents. We've done everything the doctors have told us to do, and still nothing."

What mindsets and related behavior might be challenged here?

_____

_____

_____

What would some helpful new perspectives look like?

_____

_____

_____

What actions might such a perspective lead to?

_____

_____

_____

3. Len, married with three teenage children, lost all the family's savings in a gambling spree. Under a lot of pressure from both his family and his boss, he started attending Gamblers Anonymous meetings. He

seemed to recover, that is, he stopped betting on horses and ball games. A couple of years went by and Len stopped going to the GA meetings because he "no longer needed them." He even got a better job and was recovering financially quite well. But his wife began to notice that he was on the phone a great deal with his broker. She confronted him and he reluctantly agreed to a session with one of the friends he made at GA. He says, "She's worried that I'm gambling again. You know, it's really just the opposite. When I was gambling I was financially irresponsible. I bet our future on horses and ball games. But now I'm taking a very active part in creating our financial future. Every financial planner will tell you that investing in the market is central to sound financial planning. I'm a doer. I'm taking a very active role."

What mindsets and related behavior might be challenged here?

_____

_____

_____

What would some helpful new perspectives look like?

_____

_____

_____

What actions might such a perspective lead to?

_____

_____

_____

4. Stefan, a refugee who had been brutalized by a political regime in his native country, has been seeing a counselor in a center that specializes in helping victims of torture. He has been in the country for two years. He has a decent job. Even though he is now secure, at least in some sense of that term, he has found it very difficult to establish relationships with other refugees, with members of the immigrant community, or with native-born Americans. He intimates that he is quite lonely, but dismisses loneliness as "inconsequential" when the counselor brings it up. In one session, he says: "Let's not talk about my so-called loneliness when there is a great world loneliness out there. The loneliness that brutality creates, now that's something that counts. I fill my days quite well. When it is time, I'll seek out others, but I need people who can see the world as I see it. As it really is."

What mindsets and related behavior might be challenged here?

_____

_____

_____

What would some helpful new perspectives look like?

_____

_____

What actions might such a perspective lead to?

_____

_____

_____

5. A man, 25, is talking with a fellow trainee outside the group: "In yesterday's group session I was watching Leonard [the trainer] give Peggy feedback. He does it very well. But I was also saying to myself, 'Why isn't he that helpful and that careful with me?' I want the same kind of feedback, but he doesn't say much to me at all. I'm as active as anyone else in the group. I volunteer for the demonstrations in front of the group. I don't know why he passes me by. Or is all of this just my imagination?"

What mindsets and related behavior might be challenged here?

_____

_____

_____

What would some helpful new perspectives look like?

_____

_____

_____

What actions might such a perspective lead to?

_____

_____

Chapter 11

THE SKILLS OF EFFECTIVE CHALLENGING

Specific challenging skills, as outlined in Chapter Eleven of *The Skilled Helper*, include sharing advanced empathic highlights, information sharing, helper self-disclosure, immediacy, suggestions, recommendations, confrontation, and encouragement. The purpose of these skills is to help clients get in touch with their blind spots and develop the kind of new perspectives or behavioral insights needed to complete the clarification of a problem situation or opportunity and to move on to developing possibilities for a better future, setting problem-managing goals, developing strategies, and moving to action. That is, challenging skills serve the entire helping process.

SHARING ADVANCED EMPATHIC HIGHLIGHTS

Advanced empathy, described most simply, means sharing educated **hunches** about clients and their overt and covert experiences, behaviors, and feelings that you think will help them see their problems and concerns more clearly and help them move on to developing new scenarios, setting goals, and acting. Such hunches, of course, must be based on the helper's interactions with the client in which active listening, understanding, empathy, and probing play a large part.

- Hunches can help clients see the **bigger picture**. In the following example, the counselor is talking to a client who is at odds with his brother-in-law: "The problem doesn't seem to be just your attitude toward your brother-in-law any more. Your resentment seems to be spreading to his friends. How do you see it?"
- Hunches can help clients see what they are expressing **indirectly** or merely **implying**. For example, a counselor is talking to a client who feels that a friend has let her down: "I think I might also be hearing you say that you are more than disappointed—perhaps a bit hurt and angry."
- Hunches can help clients draw **logical conclusions** from what they are saying. A manager is counseling one of his team members: "From all that you've said about her, it seems that you are also saying that right now you resent having to work with her at all. I know you haven't said that directly. But I'm wondering if you are feeling something like this."
- Hunches can help clients open up areas they are only **hinting** at. A school counselor is talking to a senior in high school: "You've brought up sexual matters a number of times, but you haven't pursued them. My guess is that sex is a pretty important area for you but perhaps pretty touchy, too."
- Hunches can help clients see things they may be **overlooking**. A counselor is talking to a man having problems at work: "I wonder if it's possible that some people take your wit too personally, that they see it, perhaps, as sarcasm rather than humor."
- Hunches can help clients identify **themes**. For instance, a counselor is talking to a woman who has been abused by her husband: "If I'm not mistaken, you've mentioned in two or three different ways that it is sometimes difficult for you to stick up for your own legitimate rights. For instance . . . ."
- Hunches can help clients **own** more fully partially-owned experiences, behaviors, and/or feelings. Example: "You sound as if you have already decided to quit."

Hunches should be based on your experience of your clients—what is happening in their lives, their behaviors and emotions, their points of view, the decisions they have made or are making, and their intended courses of action—both within the helping sessions themselves and in their day-to-day lives. Base you hunches on your relationship with your clients. Be hesitant to base your hunches on "deep" and speculative psychological theories. Be aware of the experiential and behavioral clues on which your hunches are based.

EXERCISE 47: ADVANCED ACCURATE EMPATHY—HUNCHES ABOUT YOURSELF

One way to get an experiential feeling for advanced empathy is to explore *at two levels* some situation or issue in your own life that you would like to understand more clearly. One level of understanding could be called the surface or immediate level. The second could be called a deeper or more thoughtful level. Often enough reality is found at the more thoughtful level.

1.   Read the following examples.

Example 1: A man, 25, in a counselor training program is having his ups and downs in the training group. He is having second thoughts about his ability and willingness to get close to others.

> **Level-1 understanding:** "I like people and I show this by my willingness to work hard with them. For instance, in this group I see myself as a hard worker. I listen to others carefully and I try to respond carefully. I see myself as a very active member of this group. I take the initiative in contacting others. I like working with the people here."

> **Below-the-surface understanding**: "If I look closer at what I'm doing here, I realize that underneath my hardworking and competent exterior, I am uncomfortable. I come to these sessions with more misgivings than I have admitted, even to myself. My hunch is that I have some hesitation about human closeness. I am afraid, both here and in a couple of relationships outside the group, that someone is going to ask me for more than I want to give. This keeps me on edge here. It keeps me on edge in a couple of relationships outside."

Example 2: A woman, 33, in a counselor training group, sees that her experience in the group is making her explore her attitude toward herself. It might not be as positive as she thought. She sees this as something that could interfere with her effectiveness as a counselor.

> **Level-1 understanding:** "I like myself. I base this on the fact that I seem to relate freely to others. There are a number of things I like specifically about myself. I'm fairly bright. And I think I can use my intelligence to work with others as a helper. I'm demanding of myself, but I don't place unreasonable demands on others."

> **Below-the-surface understanding**: "If I look more closely at myself, I see that when I work hard it is because I feel I have to. My hunch is that I-have-to counts more in my motivation than I-want-to. I get pleasure out of working hard, but it also keeps me from feeling guilty. If I don't work hard enough as I define the term, then I can feel guilty or down on myself. I am beginning to feel that there is too much of the I-must-be-a-perfect-person in me. This leads to my judging both myself and others more harshly than I care to think."

2.   Choose some issue, topic, situation, or relationship that you have been investigating and which you would like to understand more fully with a view to taking some kind of action on it. Try to choose issues that might affect the quality of your counseling.

3.   Briefly describe the issue, as in the examples.

4.   Then give a "surface-level" description of the issue.

5.   Next, share some hunch you have about yourself that relates to that issue, a hunch that "goes below the surface," as it were. This will help you get in touch with possible blind spots. Try to develop a new perspective on yourself and that issue, one that might help you see the issue more clearly so that you might begin to think of how you might act on it.

6.   Share you insights with your learning partner. Discuss the action implications of these insights.

Issue # 1.

_____

_____

Level-1 understanding.

_____

_____

_____

_____

Below-the-surface understanding.

_____

_____

_____

_____

_____

_____

Issue # 2.

_____

_____

Level-1 understanding.

_____

_____

_____

_____

Below-the-surface understanding.

_____

_____

_____

_____

_____

_____

EXERCISE 48: ADVANCED VERSUS BASIC EMPATHIC HIGHLIGHTS

In this exercise, assume that the helper and the client have established a good working relationship, that the client's concerns have been explored from his or her perspective, and that the client needs to be challenged to see the problem situation from some new perspective or to act on the insights that have been developed.

1. In each instance, imagine the client speaking directly to you.
2. First respond to what the client has just said with a basic empathic highlight.
3. Next, formulate one or two hunches about this person's experiences, behaviors, feelings, point of view, decision, or intended course of action—hunches that, when shared, would help promote the kind of new perspectives that drive problem-managing or opportunity-developing action. Use the client's statement and context to formulate your hunches. Ask yourself: "On what clues am I basing this hunch?"
4. Finally, respond with some form of advanced empathy, that is, share some hunch that you believe will be useful for him or her.

Example: A man, 48, husband and father, is exploring the poor relationships he has with his wife and children. In general, he feels that he is the victim, that his family is not treating him right (that is, like many clients, he emphasizes his experience rather than his behavior). He has not yet examined the implications of the ways he behaves toward his family. At this point he is talking about his pride in being a breadwinner. He says: "I work really hard to be a good provider. In fact, I'm more than a provider. I travel a lot and don't really like traveling. But this means that I can provide not just the basics but little luxuries. You know, nice holidays and things like that. But I think I'm more appreciated at work than at home. They see how hard I have to work. Often enough I have to spend hours in my home-office to prepare for my trips. For all this I get a pretty grumpy group at times."

> **Basic empathic highlight.** "It's irritating when your own family doesn't seem to appreciate what you're doing for them, the sacrifices you're making to give them a better life."
>
> **Hunch.** The family wants a husband and father, not just a breadwinner or gift-giver. He has unilaterally made a decision to work long hours in order to be the kind of provider he wants to be.
>
> **Advanced empathic highlight.** "It might be possible that their 'grumpiness' is not a sign of ingratitude but frustration because you've chosen not to be around very much."

Now try your hand at the following cases.

1. Clayton, a first-year engineering graduate student has been exploring his disappointment with himself and with his performance in school. His father is a successful engineer, but has not pressured his son to follow in his footsteps. Clayton has explored with his counselor such issues as his dislike for the school and for some of the teachers. He says: "I just don't have much enthusiasm. My grades are just okay, maybe even a little below par. I know I could do better if I wanted to. I don't know why my disappointment with the school and some of the faculty members can get to me so much. It's not like me. Ever since I can remember—even in primary school, when I didn't have any idea what an engineer was— I've wanted to be an engineer. Theoretically, I should be as happy as a lark because I'm in a graduate school with a good reputation, but I'm not."

Basic empathic highlight.

It sounds like you feel frustrated because you don't like your school and teachers, especially since your grades are lower than you expected.

its hard to accept your not doing well in school because of the school and classes, especially since you're wanted to be an engineer since you were young.

Your hunch and your reason for it.

Clayton may be resenting the school + teachers + getting poor grades because he really doesn't want to be an engineer. He maybe think his father expects it.

Advanced empathic highlight.

It must be frustrating to have sure this for however the mag choice. It might be possible that your resenting the school and teachers because you really don't want to be an engineer thinks he father expectations ✳

2. This man, now 64, retired early from work when he was 60. He and his wife wanted to take full advantage of the "golden" years. But his wife died a year after he retired. At the urging of friends he has finally come to a counselor. He has been exploring some of the problems his retirement has created for him. His two married sons live with their families in other cities. In the counseling sessions he has been alternately dealing with the theme of loss and the theme of redefining his golden years. He says: "I seldom see the kids. I enjoy them and their families a lot when they do come. I get along real well with their wives. But, since my wife has been gone, I don't make the effort I should to make it happen. I have a standing invitation from the boys and just recently I've decided I'm going to get off my sofa and start living again. I won't kid you; it will be bittersweet. I dread those times when I'll want to turn to her and enjoy the moment and she won't be there. I don't want my boys to see their father shattered and I sure as hell don't want to see pity in their eyes. I've been stuck . . . until now, at least that's what I hope."

*lonely
grieving
depressed.*

Basic empathic highlight.

your life

It's hard to re-define living after leaving a job, and having a spouse die

Your hunch and your reason for it.

possibly   grieving

Client is lonely, ~~depressed~~, since wife died + retired from work wants to connect more with sons/family but fears being vulnerable and missing his wife.

Advanced empathic highlight.

It might be possible that your afraid to get back into your life because everything has changed, and you feel vulnerable.

your ~~life~~

3. A single woman, 33, is talking to a psychiatrist about the quality of her social life. She has a very close friend, Ruth, on whom she has become somewhat dependent. She is exploring the ups and downs of this relationship. This is the third session. During the sessions she comes on a bit loud and somewhat aggressive. She says: "Ruth and I are on again off again with each other lately. When we're on, it's great. We have lunch together, go shopping, all that kind of stuff. But sometimes she seems to click off. You know, she tries to avoid me. But that's not easy to do. (She laughs.) I keep after her. She's been pretty elusive for about two weeks now. I don't know why she runs away like this. Something must be bothering her. I know we have our differences. But we always get over them."

Basic empathic highlight.

betrayed

It's frustrating when friends avoid ~~us~~ *you* *and don't do what we want them too*

Your hunch and your reason for it.

Client is draining her friend because she may be afraid to meet new people, and have more friends

91

Advanced empathic highlight.

It might be possible that your friend needs to spend time by herself or with other friends, and you may need to meet new friends too

4. A divorced woman, 35, with a daughter, 16, is talking to a counselor about her current relationship with men. She mentions that she has lied to her daughter about her sex life. She told her that she doesn't have sexual relations with men, but she does. In general she seems quite protective of her daughter. She does not know for sure if her daughter is sexually active but she has the feeling that the day is not far off when she will start having sex. She says, "I guess I've been afraid that if I told her I was sexually involved that I would lose my authority. How can I tell her to wait until she's married when I'm having sex outside of marriage. And, if I were honest how much would I have to tell her. Wait. Maybe I can be more honest with her about what I believe without needing to detail my own life. Really this is about how much I love her, not a tell-all TV show. I've wanted to connect with her on this and I think this might be the way. Sometimes, though, it feels like such a risk. What if it goes wrong?"

Basic empathic highlight.

It's hard to decide when and how to discuss sexuality with your daughter

Your hunch and your reason for it.

client is ashamed of her sexual relations and afraid to talk with her daughter about sexuality because of her own relations with men.

Advanced empathic highlight.

It may be possible that you can have an honest discussion about sexuality with your daughter and have healthy boundaries also

5. A nun, 44, a member of a counselor training group, has been talking about her dissatisfaction with her present job. Although a nurse, she is presently teaching in a primary school because, she says, of the "urgent needs" of that school. When pressed, she refers briefly to a history of job dissatisfaction. In the group she has shown herself to be an active, intelligent, and caring woman who tends to speak and act in self-effacing ways. She mentions how obedience has been stressed throughout her years in the religious order. She does mention, however, that things have been "letting up a bit" in recent years. The younger sisters don't seem to be as preoccupied with obedience as she is. She says: "The reason I'm talking about my job is that I don't want to become a counselor and then discover it's another job I'm dissatisfied with.

It would be unfair to the people I'd be working with and unfair to my religious order, which is paying for my education. Of course, I have no iron-clad assurance that I'll be put in a job that will enable me to use my counselor training."

Basic empathic highlight.

It sounds like your afraid to end up in another job/career you don't like

Your hunch and your reason for it.

Advanced empathic highlight.

It might be possible that your frustrated being your co-worker attitude about obedience has changing/changed

6. Bernard, 58, an associate professor in a university, is talking to a friend, who happens to be a counselor, about "running out of steam" in his current job. His wife works as an office manager for a large company and earns more money than he does. At one point he says, "If I were full professor, we'd be making just about the same money. But I'm not going to get promoted now." Later he says: "Well, maybe I should just try to get early retirement. We could live easily on her salary and my modest pension. Then I could work full-time at being retired (laughs). I could stumble around home just as well as stumbling around here" (laughs again). Later in the conversation he asks, "Now you, Ben, you're a bit younger than I am, but how are you going to handle 'the remains of the day' when it happens?"

Basic empathic highlight.

Its hard when your frustrated with your job and afraid you can't move up.

Your hunch and your reason for it.

Client is upset because his wife makes more money and he feels stuck in his career position

Advanced empathic highlight.

It may be possible that your frustrated because your wife makes more money and you ~~feel~~ don't think there is a chance for promotion

INFORMATION SHARING

As noted in the text, sometimes clients do not get a clear picture of a problem situation or manage it effectively because they lack information needed for clarity and action. Information can provide clients with some of the new perspectives they need to see problem situations as manageable or to spot and embrace unused opportunities. Giving clients problem-clarifying information or, better, helping them find it themselves, is not, of course, the same as advice giving or preaching.

EXERCISE 49: INFORMATION LEADING TO NEW PERSPECTIVES AND ACTION

In this exercise you are asked to consider what kind of information you might give or help clients obtain which would enable them see their problem situations or opportunities more clearly. Information can, somewhat artificially, be divided into two kinds: (a) information that helps clients understand their difficulties better and (b) information about what actions they might take.

Example: Tim was a bright, personable young man. During college he was hospitalized after taking a drug overdose during a bout of depression. He spent six months as an in-patient. He was assigned to "milieu therapy," an amorphous mixture of work and recreation designed more to keep patients busy than to help them grapple with their problems and engage in constructive change. He was given drugs for his depression, seen occasionally by a psychiatrist, and assigned to a therapy group that proved to be quite aimless. After leaving the hospital, his confidence shattered, he left college and got involved with a variety of low-paying, part-time jobs. He finally finished college by going to night school, but he avoided full-time jobs for fear of being asked about his past. Buried inside him was the thought, "I have this terrible secret that I have to keep from everyone." A friend talked him into taking a college-sponsored communication-skills course one summer. The psychologist running the program, noting Tim's rather substantial natural talents together with his self-effacing ways, remarked to him one day, "I wonder what kind of ambitions you have." In an instant Tim realized that he had buried all thoughts of ambition. After all, he didn't "deserve" to be ambitious. Tim, instinctively trusting him, divulged the "terrible secret" about his hospitalization for the first time.

Tim and the counselor, Rick, had a number of meetings over the course of a year. In some of the early sessions Rick provided him with information that proved challenging in a number of ways. For instance, when it came to employment, the counselor helped Tim find out that:

- he was more intelligent and talented than he realized and, therefore, that he was underemployed;
- that his hospitalization was quite different from, say, a felony conviction;
- deep background checks were not required for the kinds of jobs that Tim might apply for;

- privacy laws often protect prospective employees from damaging disclosures;
- enlightened employers look benignly on the developmental crises during a prospect's adolescence;
- the job market was strong and growing stronger so that people with Tim's skills and potential were in high demand.

All this challenged Tim to look at himself in a different light, that is, as a potential winner instead of a loser, and to move more aggressively into job hunting.

In the following cases you are not asked to be an information expert. Rather, from a lay person's or a common-sense point of view, you are asked to indicate what kinds of information you believe might help the client understand and manage the problem situation more clearly. What blind spots do you think the client might have? How do you think information might help? In what ways would the information be challenging?

1. Point out the blind spots the client might have.
2. Indicate what information could help the client get some new perspectives and move to action.

1. A man, 26, has just been sentenced to five years in a penitentiary on a felony burglary charge. He is talking to a chaplain-counselor who has worked for the past ten years at the penitentiary to which the man has been assigned. The man is somewhat cavalier about spending time in the penitentiary. He spends a lot of time looking for ways to overturn his conviction and is preoccupied with the possibility of early parole. His main regret is that he was caught.

What blind spots might the client have?

he is spending more time trying to get out versus changing his attitude / behavior

What information might help this client develop new perspectives?

In what way is might this information be challenging for the client?

2. A woman, 58, has endured an almost yearlong struggle to find out what has been sapping her energy. Her doctors first thought she was depressed and started her on medication but she saw little improvement and she had to contend with the side effects. Time passed and she lived in uncertainty. Recently, her doctors arrived at a diagnosis—chronic fatigue syndrome. When they told her, the woman's reaction took them a little by surprise. While she was full of questions about what it all meant, she seemed relieved and eager to get on with whatever would be required. She said it was such a relief to finally know where she stood.

What blind spots might the client have?

_____

_____

_____

What information might help this client develop new perspectives?

_____

_____

_____

_____

In what way is might this information be challenging for the client?

_____

_____

_____

_____

3. A couple in their late 50s is asked by friends to babysit for their foster children while they visit a sick parent. A one-night babysitting engagement turns into a three-day experience when an emergency prevents their friends from returning. The couple, empty nesters at this point in their lives, finds the three days both tiring and satisfying. In the days that follow they talk endlessly about what it was like to care for children again. They acknowledge that they haven't felt that alive in years. At the same time they are aware that this was just for three days. Taking in a foster child is a much larger commitment. They take these thoughts to a introductory meeting of potential foster parents sponsored by a local church agency. Later in the meeting they ask to speak with the foster care placement counselor.

What blind spots might the client have?

_____

_____

_____

_____

What information might help this client develop new perspectives?

_____

_____

_____

_____

In what way is might this information be challenging for the client?

_____

_____

_____

_____

4. A woman, 28, has been raped and is talking to a counselor at a counseling center for victims of rape. The rape took place two days ago. She has not yet reported the rape to the police. She is terrified and says that she wants to make as little fuss as possible. She feels demeaned and guilty. She fears any involvement with the police. Her instinct is to tell no one in her family.

What blind spots might the client have?

_____

_____

_____

What information might help this client develop new perspectives?

_____

_____

_____

_____

In what way is might this information be challenging for the client?

_____

_____

_____

_____

5. Tim, 18, has been smoking marijuana for about three years. He is a fairly heavy user. He has recently received several shocks. His father died suddenly and his steady girlfriend left him. Since he was not especially close to his father, he is surprised by how hard he is hit by the loss. After reading a couple of

articles on marijuana use, he developed fears that he has been doing irreversible genetic damage to himself. His is fearful of giving it up because he thinks he needs it to carry him over this period of special stress and because he fears withdrawal symptoms.

What blind spots might the client have?

_____

_____

_____

What information might help this client develop new perspectives?

_____

_____

_____

_____

In what way is might this information be challenging for the client?

_____

_____

_____

CHALLENGING NORMATIVE DISHONESTIES

As indicated in Chapter Eleven of *The Skilled Helper*, most of us face a variety of self-defeating dishonesties in our lives. We all allow ourselves, to a greater or lesser extent, to become victims of our own prejudices, smoke screens, distortions, and self-deceptions.

EXERCISE 50: CHALLENGING NORMATIVE DISHONESTIES IN ONE'S OWN LIFE

In this exercise you are asked to confront some of the ordinary ways in which we can be dishonest with ourselves and others. You are asked, as usual, to focus on the kind of dishonesties that might affect the quality of your helping or the quality of your membership in the training group.

Example 1. In this example a trainee confronts his need to control situations. She says, "I am very controlling in my relationships with others. For instance, in social situations I manipulate people into doing what I want to do. I do this as subtly as possible. I find out what everyone wants to do and then I use one against the other and gentle persuasion to steer people in the direction in which I want to go. In the training sessions I try to get people to talk about problems that are of interest to me. I even use empathic highlights and probes to steer people in directions I might find interesting. All this is so much a part of my style that usually I don't even notice it. I see this as selfish, but yet I experience little guilt about it."

Example 2. This trainee confronts his need for approval from others. He says, "Most people see me as a 'nice' person. Part of this I like, part of it is a smoke screen. Being nice is my best defense against harshness and criticism from others. I'm cooperative. I compliment others easily. I'm not cynical or sarcastic. I've gotten to enjoy this kind of being 'nice.' I find it rewarding. But it also means that I seldom talk about ideas that might offend others. My feedback to others in the group is almost always positive. I let others give feedback on mistakes. Outside the group I steer clear of controversial conversations. But I'm beginning to feel very bland."

Now confront two of your own "dishonesties," which, if dealt with, will help you be a more effective trainee and counselor. In the description section be as specific as you can. As in the examples, describe specific experiences, behaviors, feelings, points of view, and decisions. Give brief examples.

Issue # 1.

_____

_____

Describe your "dishonesties."

_____

_____

_____

_____

_____

_____

What should you do about this?

_____

_____

_____

_____

Issue # 2.

_____

_____

Describe your "dishonesties."

_____

_____

_____

_____

_____

_____

_____

What should you do about this?

_____

_____

_____

_____

HELPER SELF-DISCLOSURE

Although helpers should be **ready** to make disclosures about themselves that would help their clients understand their problem situations more clearly, they should do so only if such disclosures do not upset their clients or distract their clients from the work they are doing. Read the text on helper self-disclosure before doing this exercise.

EXERCISE 51: EXPERIENCES OF MINE THAT MIGHT BE HELPFUL TO OTHERS

In this exercise you are asked to review some problems in living or unused opportunities that you feel you have managed or are managing successfully. Indicate what you might share about yourself that would help a client with a similar problem situation. That is, what might you share of yourself that would help the client move forward in the problem-managing process?

Example 1: "In the past I have been an expert in feeling sorry for myself whenever I had to face any kind of difficulty. I know very well the rewards of seeing myself as victim. I used to fantasize myself as victim as a form of daydreaming or recreation. I think many clients get mired down in their problems because they let themselves feel sorry for themselves the way I did. I think I can spot this tendency in others. When I see this happening, I think I could share brief examples from my own experience and then ask clients to see if what I was doing squares with what they see themselves doing now."

Example 2: "I have been addicted to a number of things in my life and I see a common pattern in different kinds of addiction. For instance, I have been addicted to alcohol, to cigarettes, and to sleeping pills. I have also been addicted to people. By this I mean that at times in my life I have been a very dependent person and I found the same kind of symptoms in dependency that I did in addiction. I know a lot about the fear of letting go and the pain of withdrawal. I think I could share some of this in ways that would not accuse or frighten clients or distract them from their own concerns."

1.   List two areas in which you feel you have something to share that might help clients who have problems in living similar to your own. Just briefly indicate the area.
2.   Then indicate what you might share.
3.   Share these with a learning partner. Give and get feedback on the appropriateness and usefulness of the disclosures. Discuss the conditions under which you would share what you have written.

Area # 1.

_____

_____

What in this area might you share, if appropriate?

_____

_____

_____

_____

_____

Area # 2.

_____

_____

What in this area might you share, if appropriate?

_____

_____

_____

_____

_____

IMMEDIACY: EXPLORING RELATIONSHIPS

As noted in the text, your ability to deal directly with what is happening between you and your clients in the helping sessions themselves is an important skill. **Relationship** immediacy refers to your ability to review the history and present status of your relationship with another person, including clients and the members of your training group. **Here-and-now** immediacy refers to your ability to deal with a particular situation that is affecting the ways in which you and another person are relating right now, at this very moment.

Immediacy is a complex skill. It involves: (1) revealing how you are being affected by the other person, (2) exploring your own behavior toward the other person and its impact, (3) sharing hunches about your behavior toward him or her, (4) sharing hunches about his or her behavior toward you, (5) pointing out your own and the other's discrepancies, distortions, smoke screens, and the like, and (6) inviting the other person to explore the relationship with a view to developing a better working relationship—all through dialogue. For instance, if you see that a client is manifesting hostility toward you in subtle, hard-to-get-at ways, you may: (1) let the client know how you are being affected by what is happening in the relationship (that is, you share your experience), (2) explore how you might be contributing to the difficulty, (3) describe the client's behavior and share reasonable hunches about what is happening (challenge), and (4) invite the client to examine in a direct way what is happening in the relationship. Immediacy involves collaborative problem solving and opportunity development with respect to the relationship itself.

EXERCISE 52: RESPONDING TO SITUATIONS CALLING FOR IMMEDIACY

In this exercise a number of client-helper situations calling for some kind of immediacy on your part are described. You are asked to consider each situation and respond with some statement of immediacy. Consider the following example.

Example

- **Situation.** This client, a man of 44, occasionally makes snide remarks either about the helping profession itself or some of the things that you do in your interactions with him. At times he is cooperative and takes responsibility for himself, but at other times he has an edge and asks you to make decisions for him. He is not beyond comments such as, "Just tell me the best way for me to tell my boss that he's an idiot." He makes remarks about you personally at times, "I bet you've got a lot of friends." Or, on being invited to challenge himself, "You just won't let me alone, will you?"

- **Immediacy response.** "Tom, let's stop a minute and explore what's happening between you and me in our sessions . . . (Tom says, "Uh, oh, here we go!") . . . You take mild swipes at the counseling profession such as, 'I hear people are still trying to find out whether counseling works.' Or at me, like, 'Oh, oh, Mr. Counselor is getting a little hot under the collar.' I've ignored these remarks, but in ignoring them I've become a kind of accomplice. Sometimes we seem to be working like a team. And we get things done. Other times you ask me to make decisions for you—you know, the 'just-tell-me' approach. I've let myself get on edge with you and that's not helping us at all. . . . (Tom says, "You want to call my game, huh?") . . . Tom, I guess that I prefer that we not even play games with each other at all. Perhaps we could spend a little time re-setting our relationship."

1.   Critique the approach this counselor takes in the example. What would you change?
2.   Share your critique with a learning partner.
3.   Consider the following situations and write out an immediacy response that helps the client challenge unhelpful perspectives and actions.

102

4.   Share your responses with a learning partner and give feedback to each other. Together come up with more effective immediacy responses.

1. **The situation.** The client is a person of the opposite sex. You have had several sessions with this client. It has become evident that the person is attracted to you and has begun to make thinly disguised overtures for more intimacy. The person finds you both socially and sexually attractive. Some of the overtures have mild sexual overtones.
Immediacy response.

_____

_____

_____

_____

2. **The situation.** In the first session you and the client, a relatively successful businessman, 40, have discussed the issue of fees. At that time you mentioned that it is difficult for you to talk about money, but you finally settled on a fee at the modest end of the going rates. He told you that he thought that the fee was "more than fair." However, during the last few sessions he has been dropping hints about how expensive this venture is proving to be. He talks about getting finished as quickly as possible and intimates that is your responsibility. You, who thought that the money issue had been resolved, find it still very much alive.
Immediacy response.

_____

_____

_____

_____

3. **The situation.** The client is a young woman, 22, who is obliged to see you as part of being put on probation for a crime she committed. She is cooperative for a session or two and then becomes quite resistant. Her resistance takes the form of both subtle and not too subtle questioning of your competence, questioning the value of this kind of helping, coming late for sessions, and generally treating you like an unnecessary burden.
Immediacy response.

_____

_____

_____

_____

4. **The situation.** You are a woman. The client, 19, reminds you of your own son, 17, toward whom you have mixed feelings as he struggles to establish some kind of reasonable independence from you. The client at times acts in very dependent ways toward you, telling you that he is glad that you are helping him, asking your advice, and in various ways taking a "little boy" posture toward you. At other times he

seems to wish that he didn't have anything to do with you at all and accuses you of being "like his mother."

Immediacy response.

_____

_____

_____

_____

EXERCISE 53: IMMEDIACY WITH THE OTHER MEMBERS OF YOUR TRAINING GROUP

This exercise applies to members of an experiential training group. It could be called an immediacy "round robin." The exercise assumes that you have interacted for a number of weeks and have gotten to know one another fairly well.

1.   If the group is large, divide up into subgroups of about four trainees per group.
2.   Read the example below.
3.   On separate paper, write out a statement of immediacy for each of the members of your group. Imagine yourself in a face-to-face situation with each member successively. Deal with real issues pertaining to the training sessions, interactional style, and so forth.
4.   In a round robin, share with each of the other members of the group the statement you have written for him or her. Obviously it would be better to speak it than read it.
5.   The person listening to the immediacy statement should reply with an empathic highlight, making sure that he or she has heard the statement correctly.
6.   Then listen to the immediacy statement the other person has for you and then reply with an empathic highlight.
7.   Finally, engage in a dialogue for a few minutes about the implications of the two statements for both your interpersonal style and for this relationship.
8.   Continue with the round robin until each person has had the opportunity to share an immediacy statement and have a dialogue with every other member.

Example: Trainee A to Trainee B. "I notice that you and I have relatively little interaction in the group. You give me little feedback; I give you little feedback. It's almost as if there is some kind of conspiracy of non-interaction between us. I appreciate the way you participate in the group. For instance, the way you invite others to challenge themselves. You do it carefully but without any apology. I think I refrain from giving you feedback, at least negative feedback, because I don't want to alienate you. I do little to make contact with you. I have a hunch that you'd like to talk to me more than you do, but it's just a hunch. I'd like to hear your side of our story."

With a learning partner, identify the elements of immediacy (self-disclosure, challenge, invitation) in this example. Then move on to the exercise.

Chapter 12

THE WISDOM OF CHALLENGING

The way in which you challenge clients is extremely important. You need to challenge others in such a way that they will *respond* rather than *react* to your invitations. Before doing the following exercise, read Chapter Twelve in *The Skilled Helper*.

EXERCISE 54: EFFECTIVE VERSUS INEFFECTIVE CHALLENGE

In this exercise you are asked to practice on yourself. It goes without saying that you should not practice ineffective challenging on others, even colleagues in your training group.

1. Choose an area in which you feel that you need to challenge yourself. It can relate either to a problem situation or some unused resource or opportunity.
2. Write out a self-challenge statement in which you violate the principles outlined in Chapter Twelve.
3. Then write a self-challenge statement that embodies these principles.
4. Share both forms of self-challenge with a learning partner and discuss. Point out the flaws in the poor self-challenge. Provide feedback to each other on the quality of the appropriate self-challenge.

Example. This trainee is challenging himself on a certain lack of discipline in his life

> **Poor self-challenge**: "You're disorganized and lazy. Your room and your desk are continually messy and your excuse is that you are so busy. But that's a lame excuse. You wait until the end of the course to do the required papers and therefore their quality is poor. In some ways it is worse in your interpersonal life. You're always late. Others have to wait for you. It's a way of saying you're more important than others. You're just inconsiderate. Since all of this is so ingrained in you, you're not going to change—unless something drastic happens."

> **Effective self-challenge**: "You do reasonably well in school and your social life is quite decent. However, there is the possibility that both could be even more satisfying. At school you tend to put things off, sometimes to the point that rushing affects the quality of your work. For instance, end-of-semester papers. Since you pride yourself on giving your best, putting things off like that is not fair to yourself. You keep telling yourself it's time to get better organized, but you need to come up with a plan for doing so. What would some of the elements of such a plan be?"

EXERCISE 55: CHALLENGING ONE'S OWN STRENGTHS

Challenging unused or partially used strengths rather than weaknesses is one of the principles outlined in Chapter Twelve. In this exercise you are asked to confront yourself with respect to your own unused or underused strengths and resources.

1. Briefly identify some problem situations or unused opportunities.
2. As in the example, indicate how some unused or underused strength or strengths you have could be applied to the management of each problem situation or opportunity.
3. In addition, indicate some actions you might take to make use of underused strengths.

Example 1. Katrina, a trainee in her early 20s, is concerned about bouts of anxiety. She has led a rather sheltered life and now realizes that she needs to "break out" in a number of ways if she is to be an effective counselor. The counseling program has put her in touch with all sorts of people, and this is anxiety provoking. She lists some of the strengths she has that can help her overcome her anxiety.

> **Description of underused strengths or resources and some possible actions**: "I'm bright. I know that a great deal of my anxiety comes from fear of the unknown. I'm intellectually adventuresome. I pursue new ideas. Perhaps this can help me be more adventuresome in seeking new experiences."
>
> **Actions I can take to develop and use these resources**: "The best way to dissipate my fears is to face up to situations that cause them. New ideas don't kill me. I bet new experiences won't either. People find me easy to talk to. This can help me form friendships. My fears have kept me from developing friendships. However, developing more relationships is the best route to the kinds of different experiences I need."

Example 2. This example deals with an unused opportunity. The trainee says, "My social life is not nearly as full as I would like it to be. Extending my social life both quantitatively and qualitatively is an unused opportunity."

> **Description of underused strengths or resources**. "I have problem-solving skills, but I don't apply them to the practical problems of everyday life such as my less than adequate social life. Instead of defining goals for myself (making acquaintances, developing friendships) and then seeing how many different ways I could go about achieving these goals, I wait around to see if something will happen to make my social life fuller. I remain passive even though I have the skills to become active."
>
> **Actions I can take to develop and use resources**: "First, I am going to get active. I know people who like theater and I'll set up some evenings out. Second, there are a few others I know that are in the same boat I'm in. I think they are as ready to stop complaining about how little there is to do as I am. I'm going to bring the issue up—not to put anyone down but to engage their thinking. And, third, I'm going to make sure that when I enjoy myself on one of these evening's out that I follow-up with other social possibilities."

Now apply the same procedure to one problem situation and one unused opportunity.

Problem situation.

_____

_____

Underused strengths as applied to the problem situation.

_____

_____

_____

Actions you could take to make use of this underused resource in this problem situation.

_____

_____

_____

Unused opportunity.

_____

_____

Underused strengths as applied to the unused opportunity.

_____

_____

_____

Actions you could take to make use of this underused resource to develop this opportunity.

_____

_____

_____

EXERCISE 56: TENTATIVENESS IN THE USE OF CHALLENGING SKILLS

As noted in the text, challenges are usually more effective if they do not sound like accusations. Therefore, in challenging clients, don't accuse them, but don't be so tentative that the force of the challenge is lost.

1.   Return to Exercise 48 on advanced empathy in Chapter Eleven.
2.   With a learning partner review your responses your wrote there in terms of tentativeness.
3.   Re-do responses that would benefit from more or less or better-expressed tentativeness.
4.   Share your redone responses with a learning partner and provide mutual feedback.

EXERCISE 57: THE CHALLENGE ROUND ROBIN

The purpose of this exercise is to give you the opportunity to practice both challenging others and responding non-defensively and creatively to those who challenge you. The assumption is that you have begun to know the other members of your training group fairly well and that you are thoroughly familiar with the principles and methodology of effective challenging.

1.   Review the material on challenging and effective response to challenge in Chapters Ten, Eleven, and Twelve of *The Skilled Helper*.
2.   Divide into groups of three. The three roles are challenger, person being challenged, and observer. Challengers will do two things: (a) point out something they have noticed that their partners do well in the training group and then (b) challenge their partners in some way—for

instance, by pointing out a strength or resource that is being underused or some behavior that is dysfunctional—while following the guidelines for effective feedback.

3. The conversation should be a dialogue. In the brief exchange, the person being challenged first responds with empathic highlights to make sure that he or she understands the point of the challenge. Then he or she can use probes to further explore the point being made.

4. After the exchange, the observer gives feedback to the challenger in terms of accuracy, usefulness, and manner or style. The observer also gives feedback to the person being challenged on how well he or she responded to the challenge.

5. Each person in the triad is given a chance to play all three roles.

Example. Here is a sample dialogue.

> **Challenger.** "In our group sessions, you take pains to see to it that other members of the group are understood, especially when they talk about sensitive issues. You provide a great deal of empathy and you encourage others, principally by your example, to do the same. Your empathic highlights never sound phony and you're usually quite accurate."
>
> **Person challenged.** "You appreciate my ability and willingness to be genuinely empathic. And this adds value to group interactions."
>
> **Challenger.** "I've also noticed that you tend to limit yourself to sharing highlights. You seldom use probes and you seem to be slow to challenge anyone, for instance, by sharing hunches about the issues they are discussing. Because of your empathy and your genuineness, you have amassed a lot of credits in the group, but you don't use them to help others make reasonable demands on themselves."
>
> **Person challenged.** "So, I am less effective than I might be in that I don't move beyond empathy, even though I 'merit' doing so. I should work on increasing my 'range' within the group. I have the 'right' to challenge, but I don't use it. If I did, I'd be more helpful."

At this point the observer provides feedback.

Chapter 13

STEP I-C:
LEVERAGE: HELPING CLIENTS WORK
ON THE RIGHT THINGS

The exercises in this chapter deal with helping clients work on issues that will make a difference in their lives, that is, issues that have "leverage." Counseling time is too precious to waste on issues that do not add value and make a difference in the client's life. Read Chapter Thirteen in *The Skilled Helper* before doing the exercises in this section.

EXERCISE 58: MYSELF AS A DECISION-MAKER

Since, as a counselor, you are going to be helping clients make decisions (which, of course, is not the same as making decisions for them), it is useful to take a look at yourself as decision-maker.

1.   Read the parts of the text dealing with decision making and its shadow side.
2.   With a learning partner, discuss the difference between "rational" decision making and the "shadow side" of decision making.
3.   On your own, review the way you go about making decisions. Then, on a separate piece of paper, write a short description of your decision-making style. Include the "shadow side" of your decision-making style.
4.   Swap descriptions with a learning partner. Read them. Then discuss what you might do to improve your style.

EXERCISE 59: SCREENING PROBLEMS

Clients may need help in determining whether their issues are important enough to bring to a helper in the first place. This is called "screening."

1.   Read these two case summaries with a view to discussing them with your fellow trainees.
2.   Discuss these two cases in terms of the material in the text on screening. Consider these questions. How would you approach the woman in Case 1? Under what conditions would you be willing to work with her? How would you approach the man in Case 2? In what general ways does this case differ from Case 1? What would your concerns be in working with the man in Case 2?

Case 1: Christine got pregnant in her first year of college. She had the baby and six months later she and the child's father married. Though she left school briefly, she registered at a local college and managed to re-establish her dream of college education. The fact that she progressed a bit more slowly than she had first thought and had the responsibility of a child in her early twenties is now part of the richness in her life. She is more mature now than she would ever had been had she followed the path first laid down for her. She's certain of that. The marriage is strong, her child a blessing, and her bachelor's degree is within sight. The question now is what to do next. The career counselor at the school thinks she should consider law school. Friends tell her that she should get an M.A. in school counseling and use her experience to help others. Christine thinks of having more children, too. Fortunately, there is enough money to truly consider these options and others.

Case 2: Ray, 41, is a middle manager in a manufacturing company located in a large city. He goes to see a counselor with a somewhat complex story. He is bored with his job; his marriage is lifeless; he has poor rapport with his two teenage children, one of whom is having trouble with drugs; he is drinking heavily; his self-esteem is low; he has begun to steal things, small things, not because he needs them but because he gets a kick out of it. He tells his story in a rather disjointed way, skipping around from one problem area to another. He is a talented, personable, engaging man who seems to be adrift in life. He does not show any symptoms of severe psychiatric illness. He does experience a great deal of uneasiness in talking about himself. This is his first visit to a helper.

EXERCISE 60: CHOOSING ISSUES THAT MAKE A DIFFERENCE

Often enough, the stories clients tell are quite complex. And so they may need your help in deciding which issues to work on first and which merit substantial attention. In this exercise you are asked to review the personal concerns and problems you have identified in doing the exercises in this manual.

1.   Briefly list about ten concerns or unexploited opportunities you have discovered in doing the assessment exercises up to this point.

_____

_____

_____

_____

_____

_____

_____

_____

_____

_____

2.   Next, do some screening. Put a line through those that you would probably not bring to a counselor because they are not that important or because you believe that you could handle them easily if you wanted.

3.   Review the remaining items on your list in the light of the following "leverage" criteria:

* If there is a crisis, first help the client manage the crisis.
* Begin with the problem that seems to be causing pain for the client.
* Begin with issues the client sees as important and is motivated to work on.
* Begin with some manageable sub-problem of a larger problem situation.
* Begin with a problem that, if handled, will lead to some kind of general improvement in the client's condition.
* Focus on a problem for which the benefits will outweigh the costs.

4.	Rank order the items on your list. Then share the rankings with a learning partner. Explain the "leverage" reasons behind your rankings. Explain, in terms of the criteria outlined above or any further criteria not listed there, why you think that each has leverage. What is the "bang for the buck" of each?

Example: Gino is a trainee in a clinical psychology program. Here is one of the concerns on his list: "I am very inconsistent in the way I deal with people. For instance, some of my friends see me as fickle, I blow hot and cold. One friend told me that whenever he sees me, he's not sure which Gino he will meet. In fact, I seem to be inconsistent in other areas of life." Gino thinks this issue is important, one he has to work on. It is a question of social maturity. He also believes that if he were to handle this issue well, it would spill over into other areas of his life, because some of the same kinds of inconsistency plagues him, for instance, in his pursuit of his studies. He runs hot and cold, studying only when he wants. There is a theme here. The benefits of working on the theme will certainly outweigh the costs.

PULLING STAGE I TOGETHER

The following exercise asks you to pull together what you have learned about Stage I of the helping process and apply it to yourself.

EXERCISE 61: COUNSELING YOURSELF: AN EXERCISE IN STAGE I

At this point, you have developed an overview of the helping model and, through reading and doing the exercises in this manual, developed an understanding and behavioral "feel" for Stage I. In this exercise you are asked to carry on a dialogue with yourself in writing. Choose a problematic area of your life or some unexploited opportunity—something that is relevant to your success as a helper. First use empathic highlights, probes, and challenges to help yourself tell your story. Work at clarifying it in terms of specific experiences, behaviors, and feelings. Explore your own points of view. Clarify your decisions. Take a closer look at intended courses of action related to the issue you choose. Both probe and challenge yourself with respect to the "little" actions you can take to manage this problem or opportunity.

Example: This example comes from the experience of Cora, a woman in a master's degree program in counseling psychology. Here, then, is some of her dialogue with herself.

Her initial story. "To be frank, I have a number of misgivings about becoming a counselor. A number of things are turning me off. For instance, one of my instructors this past semester was an arrogant guy. I kept saying to myself, 'Is this what these psychology programs produce? Could this guy really help anyone?' I also find the program much too theoretical. In a *Theories of Counseling and Psychotherapy* course we never did anything, not even discuss the practical implications of these theories. And so they remained just that—theories. I'm very disappointed. I'm about to go into my second year, but I have serious reservations. From what others tell me, the program gets a bit more practical, but not enough. There's a practicum experience at the end of the program, but I want more hands-on work now. So I've started working at a halfway house for people discharged from mental hospitals. But that's not working out the way I expected either. There's something about this whole helping business that is making me think twice about myself and about the profession."

> **SELF-COUNSELOR**: All of this adds up to the fact that the helping profession, at least from your experience, is not what it's cracked up to be. The program and some of the instructors in it have left you quite disappointed. Of all these issues, which one or ones hits you the hardest?

SELF-CLIENT: Hmm. It's hard to say, but I think the halfway house bothers me most. Because that's not theoretical stuff. That's real stuff out there.

SELF-COUNSELOR: That's a place where real helping should be taking place. But you've got misgivings about what's going on there.

SELF-CLIENT: Yes, two sets of misgivings. One set about me and one about the place.

SELF-COUNSELOR: Tell me more about one set or the other.

SELF-CLIENT: I feel I have to explore both, but I'll start with myself. I felt so ill-prepared for what I encountered. What's in the lectures and books seems so distant from the realities of the halfway house. For instance, the other day one of the residents there began yelling at me when we were passing in the hallway. She hit me a few times and then ran off screaming that I was trying to get her in some way. I didn't know what to do. I felt so incompetent. And I felt guilty.

SELF-COUNSELOR: That sounds pretty upsetting. You just weren't prepared for it. I'm wondering whether the more practical part of the counseling program you're going into would better prepare you for that kind of reality.

SELF-CLIENT: It could be. I may be doing myself in by jumping ahead of myself.

SELF-COUNSELOR: But you still have reservations about the effectiveness of the halfway house.

SELF-CLIENT: I wasn't ready for what I found there. I've been there a couple of months. No one has really helped me learn the ropes. I don't have an official supervisor. I just see all sorts of people with problems and help when I can.

SELF-COUNSELOR: You just don't feel prepared and they don't do much to help you. So you feel inadequate. You also seem to be basing your judgments about the adequacy of helper-training programs and helping facilities in general, on your experience in this training program, and on the halfway house.

SELF-CLIENT: That's a good point. I'm making the assumption that both should be high-quality places. As far as I can tell, they're not. I guess I have to work on myself first. But that's why I went to the halfway house in the first place. I'm an independent person, but I'm too much on my own there. In a sense, I'm trusted, but, since I don't get much supervision, I have to go on my own instincts and I'm not sure they're always right.

SELF-COUNSELOR: There's some comfort in being trusted, but without supervision you still have a what-am-I-doing-here feeling.

SELF-CLIENT: Right. There are many times when I ask myself just that, "What are you doing here?" I provide day-to-day services for a lot of people. I listen to them. I take them places, like to the doctor. I get them to participate in conversations or games and things like that. But it seems that I'm always just meeting the needs of the moment. I'm not sure what the long-range goals of the place are and if anyone, including me, is contributing to them in any way.

SELF-COUNSELOR: You seem to have started with a rather difficult population. You get some satisfaction in providing the services you do, but you don't feel equipped to add real value in the halfway house. You may even wonder what this value might look like. Anyway, the lack of overall purpose or direction for yourself and the institution is frustrating. I'm not sure whether you've asked yourself what you might do about all of this, either at the halfway house or in the counseling program.

SELF-CLIENT: I'm letting myself get frustrated, irritated, and depressed. I'm down on myself and down on the people who run the house. It's a day-to-day operation that sometimes seems to be a fly-by-night venture. See! There I go. You're right. I'm not doing anything to handle my frustrations. I'm just blaming everyone else. And I'm doing little to try to better myself. I'm just beginning to put it into some kind of context.

SELF-COUNSELOR: So this is really the first time you've stopped to take a critical look at yourself as a potential helper and the settings in which helping takes place. And what you see is depressing.

SELF-CLIENT: Absolutely.

SELF-COUNSELOR: I wonder how fair you're being to yourself, to the profession, and even to the halfway house.

SELF-CLIENT (Pause): Well, that's a point. I'm speaking—and making judgments—out of a great deal of frustration. I don't want to go off half cocked. But I could share my concerns about the psychology program with one of the instructors. She teaches the second-year trainees. She gave a very practical talk. I could also see whether my concerns are shared by my classmates. I could also talk to the second-year students to get a feeling for how practical next year might be.

SELF-COUNSELOR: All in all, you can give yourself a chance to get a more balanced perspective.

SELF-CLIENT: Right. I need to do something instead of just brooding and complaining. But there's still the halfway house. I don't want to come across as the wet-behind-the-ears critic.

SELF-COUNSELOR: I'm not sure whether you're assuming that halfway house you're working in should be a state-of-the-art facility?

SELF-CLIENT: Touché! . . . Wow, I am basing a lot of my feelings about the profession on that place.

SELF-COUNSELOR: Who at the house might you trust to help you get a better perspective?

SELF-CLIENT: Well, hmm . . . There is a consultant who shows up once in a while. He runs staff meetings. And he does it in a very practical way. Maybe I can get hold of him and get a better picture.

SELF-COUNSELOR: So, overall?

SELF-CLIENT: I've got some work to do before I rush to judgment. I like the fact that I want the people and the institutions in the profession to be competent. Deep down I think that I'd make a good helper. I say to myself, "You're all right; you're trying to do what is right." Also, I need to challenge my idealism. I spend too much time grieving over what is happening—or not happening—at school and at the halfway house. I need a more mature outlook and I need to figure out how to turn minuses into pluses, including my own.

1. Review this trainee's responses to herself with a learning partner. What kinds of responses did she use? How would you evaluate their quality? What progress did she make? Describe the movement she made during the session. What responses would you have changed?

2. Next choose a problematic area that is important to you and on separate sheets of paper engage in the same kind of dialogue with yourself.

3. Exchange dialogues with your learning partner and engage in the same kind of critique outlined above.

PART FOUR

STAGE II:
HELPING CLIENTS DETERMINE WHAT
THEY NEED AND WANT

Stages II and III together with the Action Arrow get at the heart of

helping because they deal with solutions—goals, outcomes, and

action. Chapter 14 is an introduction to Stage II. It focuses on the

nature of solutions, the relationship between solutions and brief

therapy, the importance of goal setting, and the peculiar nature of

decision making. Chapter 15, Step II-A, deals with helping clients

review possibilities for a better future. Because it deals with the

future, it also focuses on the place of hope in helping. Chapter 16,

Step II-B, outlines a process for helping clients choose and shape

their goals. Chapter 17, Step II-C, reviews the importance of

commitment to goals and to the work clients must do to achieve

these goals. It also deals with self-efficacy and helping clients find

the incentives for moving forward.

Stage II focuses on the client's desire or need for a better future. Both problems and unused opportunities can make clients feel hemmed in and closed off. To a greater or lesser extent they have no future, or the future they have looks troubled. The steps of Stage II outline three ways in which helpers can be with their clients at the service of exploring and developing this better future.

Step II-A: Helping clients identify *possibilities* for a better future. What do you want? What do you need? What are some of the possibilities?

Step II-B: Helping clients choose the specific *goals* that will constitute a better future. Given the possibilities, what do you really want? What are your choices?
Step II-C: Helping clients discover incentives for *commitment* to their goals. What are you willing to pay for what you want?

Chapter 14

INTRODUCTION TO STAGE II: HELPING CLIENTS IDENTIFY, CHOOSE, AND SHAPE GOALS: "What Do I Need And Want?"

In many ways Stages II and III together with the Action Arrow are the most important parts of the helping model because they are about "solutions." It is here that counselors help clients develop and implement programs for constructive change. The payoff for identifying and clarifying both problem situations and unused opportunities lies in doing something about them. The skills helpers need to help clients to do precisely that—engage in constructive change—are reviewed and illustrated in Stages II and III. Read Chapter Fourteen before doing these exercises.

EXERCISE 62: THE VALUE OF GOAL SETTING

Too often we take goal setting for granted. We set goals everyday, usually without realizing that we are doing so.

1.   Picture yourself talking with a client, Bonita, who is very good at telling her story but avoids going beyond Stage I.
2.   Your task is to explain to her how important it is to move from a discussion about her problems and opportunities to managing them.
3.   On a separate piece of paper, write out what you might say to her about the value of Stages II and III. In real life, of course, you would do this through dialogue with Bonita.
4.   Share what you have written with a learning partner. Give feedback to each other on what you both have written. If you were Bonita, how might you react to what your learning partner is saying in his or her statement?

EXERCISE 63: HOW GOAL ORIENTED ARE YOU?

If you are to help clients set goals and come up with strategies for accomplishing them, it is helpful to understand your own approach to goal setting.

1. When it comes to goal setting, whether formal or informal, what are your strengths?

_____

_____

_____

_____

2. When it comes to goal setting, whether formal or informal, what are your weaknesses?

_____

_____

_____

_____

_____

3. What do you need to do to become a better goal setter in your own personal life?

_____

_____

_____

_____

_____

_____

EXERCISE 64: YOUR DECISION-MAKING STYLE

Since, as a counselor, you will be helping clients make decisions about a wide range of issues, you can benefit by taking a look a your own decision-making style. First read the section of Chapter Fourteen on helping clients become more effective decision-makers.

1. After reading the section on rational decision making, list your strengths as a "rational" decision-maker.

_____

_____

_____

_____

_____

_____

2. Next, read the section on the shadow side of decision making. In what ways and in what circumstances does your decision making end up in the shadows?

_____

_____

_____

_____

_____

_____

_____

_____

3. After reading the section on making smarter decisions, indicate what you need to do to become a more effective decision-maker.

_____

_____

_____

_____

_____

_____

Chapter 15

STEP II-A: HELPING CLIENTS CREATE A BETTER FUTURE
"What Do You Want?"—DISCOVERING POSSIBILITIES

Effective helping requires imagination. In this step you are asked to help yourself and clients develop a vision of a better future. Once clients understand the nature of the problem situation or identify a key unused opportunity, they need to ask themselves, "What do I want? What would my situation look like if it were better, at least a little bit better?" Read Chapter Fifteen in the text before doing these exercises.

EXERCISE 65: PROBLEMS AND OPPORTUNITIES IN THE SOCIAL SETTINGS OF LIFE

This is another exercise that will help you review areas in which clients have problems and also help you identify your own strengths and soft spots. Individuals belong to and participate in a number of different social settings in life: family, circle of friends, clubs, church groups, school groups, and the like. People are also affected by what goes on in our neighborhoods and the cities and towns in which we live. Larger systems, such as state and national governments, have their ways of entering people's lives.

1. **Chart the social settings of life.** Chart the social settings of your life as in the example below. The person in the example is Mitch, 45, a principal of an inner-city high school in a large city. He is married and has two teenage sons. Neither attends the high school of which he is principal. He is seeing a counselor because of exhaustion and bouts of hostility and depression. He has had a complete physical check-up and there is no evidence of any medical problem.

2. **Choose two or three social settings that are currently key**. These are settings in which you are experiencing problems or have unused opportunities.

3. **Review issues, demands, conflicts, concerns**. Now take each key social setting and write down issues or concerns in that setting. For instance, some of the things Mitch writes are:

School

- Some faculty members want a personal relationship with me and I have neither the time nor the desire.
- Some faculty members have retired on the job. I don't know what to do with them.
- Some of the white faculty members are suspicious of me and distant just because I'm black.
- One faculty member wrote the district superintendent and said that I was undermining her reputation with other faculty members. This is not true.
- The students, both individually and through their organizations, keep asking me to be more liberal while their parents are asking me to tighten things up.

Family

- My wife says that I'm letting school consume me; she complains constantly because I don't spend enough time at home. Even though she wants me to spend more time at home and yet she criticizes me for not spending more time with my parents.
- My kids seem to withdraw from me because I'm a double authority figure, a father and a principal.

Parents

- My mother is infirm; my retired father calls me and tells me what a hard time he's having getting used to retirement.
- My mother tells me not to be spending time with her when I have so much to do and then she complains to my wife and my father when I don't show up.

School District

- I would like to become the district superintendent. I think I could do a lot of good.

Friends

- My friends say that I spend so much time at work involving myself in crisis management that I have no time left for them; they tell me I'm doing myself in.

4. **Possibilities for a better future.** Choose two concerns arising from two different social settings of your life. Spell out some of the possibilities for a better future in this area. What would this problem or unused opportunity look like if it were managed better or solved? What are some of the things that would be in place that are not now in place? Use the following probes to help yourself brainstorm possibilities.

- Here's what I need . . .
- Here's what I want . . .
- Here are some items on my wish list . . .
- When I'm finished I will have . . .
- There will be . . .
- I will have in place . . .
- I will consistently be . . .
- There will be more of . . .
- There will be less of . . .

For instance, Mitch, in reviewing the conflict between his work and his friends, comes up with these possibilities: more time with his friends, greater flexibility in managing his calendar at work, better integration of his friendships with his home life, a more clear-cut division between his work setting and the other social settings of his life, more time for himself, a clearer understanding of his career aspirations and the costs associated with them, and better delegation on his part to the members of the school administrative team. Now do the same for yourself in two or three of the social settings of your life on a separate sheet of paper.

First social setting.

_____

Social setting related issue, demand, conflict, opportunity, concern.

_____

Possibilities for a better future in this area of concern.

_____

_____

_____

_____

_____

Second social setting.

_____

Social setting related issue, demand, conflict, opportunity, concern.

_____

Possibilities for a better future in this area of concern.

_____

_____

_____

_____

_____

Third social setting.

_____

Social setting related issue, demand, conflict, opportunity, concern.

_____

Possibilities for a better future in this area of concern.

_____

_____

_____

_____

_____

EXERCISE 66: BRAINSTORMING POSSIBILITIES FOR A BETTER FUTURE—CASES

This exercise focuses on a number of cases, many of which you have seen before. Now you are asked to put yourself in each person's shoes and brainstorm possibilities for a better future. What would things look like if they looked better? If you were this person, what are some of the things you would want instead of what you've got?

1.   Read the case. Put yourself in this person's shoes. Make his or her problem situation your own.
2.   Point out a blind spot this person might have—a blind spot that might keep this person from managing the problem situation or developing an opportunity more effectively.
3.   Brainstorm a number of possibilities for a better future. What would you want if you were this person?
4.   Use the probes outlined in the previous exercise to develop possibilities for a better future.

1. Tim was a bright, personable young man. During college he was hospitalized after taking a drug overdose during a bout of depression. He spent six months as an in-patient. He was assigned to "milieu therapy," an amorphous mixture of work and recreation designed more to keep patients busy than to help them grapple with their problems and engage in constructive change. He was given drugs for his depression, seen occasionally by a psychiatrist, and assigned to a therapy group that proved to be quite aimless. After leaving the hospital, his confidence shattered, he left college and got involved with a variety of low-paying, part-time jobs. He finally finished college by going to night school, but he avoided full-time jobs for fear of being asked about his past. Buried inside him was the thought, "I have this terrible secret that I have to keep from everyone." A friend talked him into taking a college-sponsored communication-skills course one summer. The psychologist running the program, noting Tim's rather substantial natural talents together with his self-effacing ways, remarked to him one day, "I wonder what kind of ambitions you have." In an instant Tim realized that he had buried all thoughts of ambition. After all, he didn't "deserve" to be ambitious. Tim, instinctively trusting the program director, divulged the "terrible secret" about his hospitalization for the first time.

A key blind spot translated into a future-oriented new perspective.

_____

_____

A range of possibilities for a better future.

_____

_____

_____

_____

_____

2. Cormack is in a master's degree program in counseling psychology. He says: "To be frank, I have a number of misgivings about becoming a counselor. A number of things are turning me off. For instance, one of my instructors this past semester was an arrogant guy. I kept saying to myself, 'Is this what these psychology programs produce? Could this guy really help anyone?' I also find the program much too theoretical. In a 'Theories of Counseling and Psychotherapy' course we never did anything, not even discuss the practical implications of these theories. And so they remained just that—theories. I'm very

disappointed. I'm about to go into my second year, but I've got serious reservations. From what others tell me, the program gets a bit more practical, but not enough. There's a practicum experience at the end of the program, but I need more hands-on work now. So I've started working at a halfway house for people discharged from mental hospitals. But that's not working out the way I expected either. There's something about this whole helping business that is making me think twice about myself and about the profession."

A key blind spot translated into a future-oriented new perspective.

_____

_____

A range of possibilities for a better future.

_____

_____

_____

_____

_____

3. A nun, 44, a member of a counselor training group, has been talking about her dissatisfaction with her present job. Although a nurse, she is presently teaching in a primary school because, she says, of the "urgent needs" of that school. When pressed, she refers briefly to a history of job dissatisfaction. In the group she has shown herself to be an active, intelligent, and caring woman who tends to speak and act in self-effacing ways. She mentions how obedience has been stressed throughout her years in the religious order. She does mention, however, that things have been "letting up a bit" in recent years. The younger sisters don't seem to be as preoccupied with obedience as she is. She says: "The reason I'm talking about my job is that I don't want to become a counselor and then discover it's another job I'm dissatisfied with. It would be unfair to the people I'd be working with and unfair to my religious order, which is paying for my education. Of course, I have no iron-clad assurance that I'll be put in a job that will enable me to use my counselor training."

A key blind spot translated into a future-oriented new perspective.

_____

_____

A range of possibilities for a better future.

_____

_____

_____

_____

_____

4. Marcus, 47, is a successful businessman whose life is in fairly good order. However, Yves, his younger brother, is struggling financially. His credit-card debt is enormous. For Marcus the problem is how to extend a helping hand to his younger brother without coming off as pitying or being seen that way. He explains his dilemma to a colleague at work, "Yves is very sensitive to this older brother-younger brother stuff. Anyway, he has it in his head that I see him as incompetent and any help I offer is seen through that filter. I suppose that I'll always be the successful big brother and he will always be the inept little brother. It's not that way really but.... What I can't do is let our past history force me to stand off to the side with my hands in my pockets and do nothing when I know I could be of some help. I'm thinking of just writing him a check and saying, "Let's just make a new beginning in every way."

A key blind spot translated into a future-oriented new perspective.

_____

_____

A range of action possibilities.

_____

_____

_____

_____

5. This man, now 64, retired early from work when he was 62. He and his wife wanted to take full advantage of the "golden" years. But his wife died a year after he retired. At the urging of friends he has finally come to a counselor. He has been exploring some of the problems his retirement has created for him. His two married sons live with their families in other cities. In the counseling sessions he has been dealing somewhat repetitiously with the theme of loss. He says: "I seldom see the kids. I enjoy them and their families a lot when they do come. I get along real well with their wives. But now that my wife is gone . . . (pause) . . . and since I've stopped working . . . (pause) . . . I seem to just ramble around the house aimlessly, which is not like me at all. I suppose I should get rid of the house, but it's filled with a lot of memories—bittersweet memories now. There were a lot of good years here."

A key blind spot translated into a future-oriented new perspective.

_____

_____

A range of possibilities for a better future.

_____

_____

_____

_____

EXERCISE 67: MY POTENTIAL AS A HELPER

You may or may not want to become either a paraprofessional or professional helper. You may be involved in this course simply to become a better helper in everyday life; no matter what career you choose to pursue. In this exercise assume that one of these three options is yours.

1.   Describe two facets of what you are currently like as a helper.
2.   Describe, in terms of possibilities, the kind of helper you would like to become. Be as concrete as possible.
3.   Share your two descriptions with a learning partner. Mutually clarify and refine these possibilities through empathic highlights and probing. After listening to the other, indicate further possibilities you might want to add to your list.

1. Here is what I am currently like.

_____

_____

_____

_____

_____

Here is what I'd like to become.

_____

_____

_____

_____

2. Here is what I am currently like.

_____

_____

_____

_____

_____

Here is what I'd like to become.

_____

_____

_____

_____

_____

EXERCISE 68: HELPING ANOTHER DEVELOP POSSIBILITIES FOR A BETTER FUTURE

In this exercise, you are asked to help one of the other members of your group develop possibilities for a better future.

1.   Divide up into groups of three for this exercise.
2.   Assume, sequentially, the roles of client, helper, and observer.
3.   In the role of client, provide a summary of one of the problem situations or undeveloped opportunities you focused on in any of the previous exercises. Or, if you prefer, choose a new issue.
4.   In the role of counselor, listen to your partner summarize his or her problem situation or undeveloped opportunity. Use empathy and probes to get a full statement of the issue.
5.   Finally, help the other develop possibilities for a better future. Use empathy, probes, summaries, and challenges to help the other expand the list.

Example: Geraldo, a junior in college majoring in business studies, has given the following summary of the problem situation: "My uncle runs a small business. He's offered me a kind of internship in the family business. While such an opportunity fits perfectly with my career plans, I have done nothing to develop it. Right now, it's only an interesting idea. But it still seems too good an opportunity to miss." Trish, his helper, uses all the communication skills learned up to this point to help him develop possibilities for a productive internship. Some of the dialogue goes like this:

> **TRISH:** So you haven't taken the time yet to spell out what the internship could look like.
> **GERALDO:** No. Generally, I see the internship as something that would bring to life the stuff I'm getting out of business books right now.
> **TRISH:** So the internship would give a chance for some hands-on experience with business realities that are just concepts now. What could the internship look like?
> **GERALDO:** Well, I don't know what's on my uncle's mind. Though he's not going to push me into anything I don't want.
> **TRISH:** Then it seems to make sense for you to have some ideas in mind before you talk with him about it. What would _you_ like the internship to look like? What do you want?
> **GERALDO:** Let's see. I'd be putting in 10 to 15 hours a week at my uncle's business. Actually, that would make sense since I'm more or less overbalanced right now on social activities.
> **TRISH:** What would you be doing during those hours?
> **GERALDO:** Well, I don't like the finance part of my program. So one thing I could do is become familiar with the finance part of his business—where the money comes from, what kind of debt he carries, cash flow—all those things that are still too theoretical for me in the courses I'm taking.

TRISH: You'd see finance in action. It might even seem more interesting to you then. What else do you picture yourself doing?

GERALDO: I like to find out what management and supervision actually looks like inside a company. I get a confused picture of both of these from what I'm reading.

TRISH: So it's one thing to read management books, but another to see how it actually works. What might you do as a manager or supervisor?

GERALDO: I'm not sure he'd let me run anything important. But maybe I could run a project or two. And the project might involve working with others—you know, like a small team. I think I'd like to see how teamwork can affect the business. I guess what I'm saying is that my uncle's business could be a lab for me. I'd be contributing, but also I'd be learning a lot.

TRISH: So project work and teamwork could be important parts of the lab. But it also sounds like you want to be a contributor in all this.

GERALDO. Yeah. I'd like to learn, but I'd like to make a difference.

TRISH. What kind of difference?

The dialogue goes on in that vein. Geraldo, with the help of Trish, develops a lot of possibilities for the internship.

6.   With your partner critique this dialogue. What did Trish do well? What might have been done better?
7.   Then move on to your own dialogue.
8.   After eight or ten minutes get feedback from the client first and then the observer as to how useful you have been in helping the client develop possibilities.
9.   Repeat the process until each person has played each role.

Chapter 16

STEP II-B: MAKING CHOICES—
"What Do You Really Want?"

Once clients have brainstormed possibilities for a better future, they need to make some choices. Answers to the questions, "What do you want?" and "What do you need?" are possible goals. Answers to the question, "What do you really need and want?" are the actual goals that constitute the change agenda. It is as if the client were to say, "Here is what I want in place of what I have right now." Goals are a critical part of an agenda for constructive change. For instance, Geraldo, the college student seen in Chapter 15, after brainstorming a number of possibilities for structuring his internship in his uncle's company, chooses four of the possibilities and makes these his agenda. He shares them with his uncle and, after some negotiation, comes up with a revised package they can both live with. Read Chapter Sixteen before doing these exercises.

EXERCISE 69: TURNING POSSIBILITIES INTO GOALS—A CASE

In this exercise you are asked to do what Geraldo had to do—choose goals from among the possibilities. Possibilities should be chosen because they will best help you manage your problem situation or develop some opportunity. First, read the following case.

The Case. Vanessa, 46, has been divorced for about a year. She has done little to restructure her life and is still in the doldrums. At the urging of a friend, she sees a counselor. With the counselor's help, she brainstorms a range of possibilities for a better future around the theme of "my new life as a single person." She is currently a salesperson in the women's apparel department of a moderately upscale store. She lives in the house that was part of the divorce settlement. She has no children. When she finally decided that she wanted children, it was too late. There was some discussion with her ex-husband about adopting a child, but it didn't get very far. The marriage was already disintegrating. Her visits to the counselor have reawakened a desire to take charge of her life and not just let it happen. She let her marriage happen and it fell apart. She is not filled with anger at her former husband. If anything, she's a bit too down on herself. Her grieving is filled with self-recrimination.

The helper asks her, "What do you want now that the divorce is final? What would you like your post-divorce life to look like? What would you want to see in place?" Vanessa brainstormed the following possibilities:

- A job in some area of the fashion industry.
- A small condo that will not need much maintenance on my part instead of the house.
- The elimination of poor-me attitudes as part of a much more creative outlook on self and life.
- The elimination of waiting around for things to happen.
- The development of a social life. For the time being, a range of friends rather than potential husbands. A couple of good women friends.
- Getting into physical shape.
- A hobby or avocation that I could get lost in. Something with substance.
- Some sort of volunteer work. With children, if possible.
- Some religion-related activities, not necessarily established-church related. Something that deals with the "bigger" questions of life.
- An end to all the self-recrimination over the divorce.
- Resetting my relationship with my mother (who strongly disapproved of the divorce).

- Getting over a deep-seated fear that my life is going to be bland, if not actually bleak.
- Possibly some involvement with politics.

1.   Read the list from Vanessa's point of view (even though you hardly know her).
2.   Since Vanessa cannot possibly do all of these at once, she has to make some choices. Put yourself in her place. Which items would you include in your agenda if you were Vanessa? In what way would the agenda items, if accomplished, help manage the overall problem situation and develop opportunities? Choose three.
3.   Give your reasons for including each.
4.   Share your package together with the reasons for each item in it with a learning partner. Discuss the differences in and the reasons for the choices. Note that there is no one right package.

My first choice.

_____

My reason for the choice.

_____

_____

_____

My second choice.

_____

My reason for the choice.

_____

_____

_____

My third choice.

_____

My reason for the choice.

_____

_____

_____

EXERCISE 70: TURNING PERSONAL POSSIBILITIES INTO PERSONAL GOALS

In this exercise you are asked to review EXERCISE 65: PROBLEMS AND OPPORTUNITIES IN THE SOCIAL SETTINGS OF LIFE in which you brainstormed a range of possibilities for a better future for yourself.

1.   Pick two of the three social-settings you explored.

2.   In each, after reviewing the possibilities that you brainstormed, come up with a goal and your reason for it.
3.   Use the criteria outlined in Chapter Sixteen in order to "fashion" or "design" the goals.
4.   After completing the exercise, share and critique your goals with a learning partner.

1. First social setting.

_____

Goal for this setting.

_____

_____

Reason for the goal.

_____

_____

_____

2. Second social-setting.

_____

Goal for this setting.

_____

_____

Reason for the goal.

_____

_____

_____

EXERCISE 71: HELPING CLIENTS TURN POSSIBILITIES INTO GOALS

In this exercise you are asked to review EXERCISE 66: BRAINSTORMING POSSIBILITIES FOR A BETTER FUTURE—CASES in which you brainstormed a range of possibilities for a better future for three different clients.

1.   Pick two of the three cases in that exercise.
2.   In each, turn the possibilities that you brainstormed into a workable goal or set of goals.
3.   Use the criteria outlined in Chapter Sixteen in order to "fashion" or "design" the goals.
4.   After completing the exercise, share and critique your goals with a learning partner.

1. First social-setting situation.

_____

_____

Possibilities turned into a goal.

_____

_____

_____

_____

Reason for the goal.

_____

_____

2. Second social-setting situation.

_____

_____

Possibilities turned into a goal.

_____

_____

_____

_____

_____

Reason for the goal.

_____

_____

EXERCISE 72: SHAPING—MAKING GOALS SPECIFIC

Many clients set goals that are too vague to be accomplished. "I'm going to clean up my act" is nothing more than a good intention. It is not yet a goal. Part of shaping, then, is making goals specific enough to drive action. This exercise involves moving from vagueness to concreteness.

1.   First, evaluate the goal Tom and Carol in the example below have set for themselves in terms of its specificity. If you don't think it is specific enough, shape it, that is make it more specific.
2.   Then, in the cases that follow, get inside the client's mind and try to determine what he or she might need or want.
3.   Finally, move from a vaguely stated *good intention* to a *broad aim* and then to a *specific goal*.
4.   Finally, learning partners should share the goals they have written and provide feedback for each other. When necessary, goals should be reformulated.

Example: Tom, 42, and his wife, Carol, 39, have been talking to a counselor about how poorly they relate to each other. They have explored their own behavior in concrete ways, have developed a variety of new perspectives on themselves as both individuals and as a couple, and now want to do something about what they have learned. In broad terms, they have agreed to stop blaming each other for their problems and take mutual responsibility for creating a better relationship. What follows is their first shot at moving from a vague to a more specific goal in the area of communication:

> **Good Intention**: "We've got to do something about the way we communicate with each other. It's just not healthy."
> **General Aim**: "We are constantly at each other over inconsequential issues and then it escalates. We'd like to do something about the way we bicker."
> **Specific Goal**: "We want to call a moratorium on what has become almost childish bickering. We want to stop second-guessing, picking on, correcting, one-upping, putting down, questioning, and being snide with each other. But we don't want to leave a vacuum. We need something positive to take its place. For instance, when we kid around, each of us should make himself or herself, not the other, the butt of any joke. When we do disagree about something important, we should avoid overly personal heated arguments and instead engage in mutual problem solving based on the skilled helping process we are using in counseling."

First, critique the specific goals in the above example and then work on the cases below.

1. A woman, 53, is talking to her lawyer-counselor. She has recently received a significant sum of money from a deceased relative. "I suppose I knew my aunt was going to remember me but I had no idea that it would be this much. I've spent most of my life managing small amounts of money over short periods of time—like until the end of the month. I have no experience with this kind of money. I could go to school. I could travel. I certainly will want to set aside some money for others in the family. I'm excited and stunned at the same time. I'm the deer in the headlights everyone talks about."

Statement of good intention.

_____

_____

Broad aim.

_____

_____

Specific goal.

_____

_____

_____

_____

_____

2. Troy, 30, has been discussing the stress he has been experiencing during this transitional year of his life. Part of the stress relates to his job. He has been working as an accountant with a large firm for the past five years. He makes a decent salary, but he is more and more dissatisfied with the kind of work he is doing. He finds accounting predictable and boring. He doesn't feel that there's much chance for advancement in this company. Many of his associates are much more ambitious and much more inventive than he is.

Statement of good intention.

_____

_____

Broad aim.

_____

_____

Specific goal.

_____

_____

_____

_____

3. A high-school senior is talking to a school counselor about college and what kinds of courses she might take there. However, she also mentions, somewhat tentatively, her disappointment in not being chosen as valedictorian of her class. She and almost everyone else had expected her to be chosen. She says: "I know that I would have liked to have been the class valedictorian, but I'm not so sure that you are supposed to count on anything like that. They chose Jane. She'll be good. She speaks well and she's very popular. But no one has a right to be valedictorian. I'd be kidding myself if I thought differently. I've done better in school than Jane, but I'm not as outgoing or popular."

Statement of good intention.

_____

_____

Broad aim.

_____

_____

Specific goal.

_____

_____

_____

_____

_____

4. Len, married with three teenage children, lost all the family's savings in a gambling spree. Under a lot of pressure from both his family and his boss, he started attending Gamblers Anonymous meetings. He seemed to recover, that is, he stopped betting on horses and ball games. A couple of years went by and Len stopped going to the GA meetings because he "no longer needed them." He even got a better job and was recovering financially quite well. But his wife began to notice that he was on the phone a great deal with his broker. She confronted him and he reluctantly agreed to a session with one of the friends he made at GA. He says, "She's worried that I'm gambling again. You know, it's really just the opposite. When I was gambling I was financially irresponsible. I lost our future on horses and ball games. But now I'm taking a very active part in creating our financial future. Every financial planner will tell you that investing in the market is central to sound financial planning. I'm a doer. I'm taking a very active role."

Statement of good intention.

_____

_____

Broad aim.

_____

_____

Specific goal.

_____

_____

_____

_____

5. Joan, 32, is married and has two small children. Her husband has left her and she has no idea where he is. She has no relatives in the city and only a few acquaintances. She is talking to a counselor in a local community center about her plight. Since her husband was the breadwinner, she now has no income and no savings on which to draw.

Statement of good intention.

_____

_____

Broad aim.

_____

_____

Specific goal.

_____

_____

_____

_____

_____

EXERCISE 73: SHAPING GOALS—MAKING SURE THAT THEY ARE WORKABLE

Accomplishing goals is a lot of work. If the goals chosen are themselves flawed, then the job becomes almost impossible. For instance, if Radcliffe chooses a career for which he does not have the talent, he is doing himself no favor. Hard work won't make up for the talent deficit. In order to be workable, goals generally need to be:

- stated as _outcomes_,
- _specific_ enough to drive action,
- _substantial_ enough to make a _difference_,
- characterized by the right mixture of _risk_ and _prudence_,
- _realistic_,
- _sustainable_,
- _flexible_,
- _congruent_ with the client's _values_,
- set in a _reasonable time frame_.

These characteristics can be seen as tools that help you shape both your own and clients' goals.

1.   With a learning partner return to the four goals that each of you set in the previous two exercises. Use the above characteristics to review them once more.
2.   What changes might you make in each of these goals to make them conform to the above characteristics?

EXERCISE 74: HELPING LEARNING PARTNERS SET WORKABLE GOALS

In this exercise you are asked to act as a helper to a learning partner. This exercise is done verbally.

1.   The total training group is to be divided up into smaller groups of three.
2.   There are three roles: client, helper, and observer. Decide the order in which you will play each role.
3.   The client summarizes some problem situation and then declares his or her **intent** to do something about the problem or some part of it.
4.   The group member in helper role, using empathic highlights, probing, and challenging, helps the client move from this statement of intent to a broad aim and then to a specific problem-managing or opportunity-developing goal that has the characteristics listed above.
5.   When the helper feels that he or she has fulfilled this task, the session is ended and both observer and client give feedback to the helper on his or her effectiveness.
6.   Repeat the process until each person has played each role.

Example: Since most students do not operate at 100% efficiency, there is usually room for improvement in the area of learning. Luisa, a junior beginning her third year of college, is dissatisfied with the way she goes about learning. She decides to use her imagination to invent a more creative approach to study. She brainstorms possibilities for a better study future, that is, goals that would constitute her new learning style. She says to herself: "In my role as student or learner, what do I want? Let me brainstorm the possibilities."

•   I will stop studying for grades and start studying to learn. Paradoxically this might help my grades, but I will not be putting in extra effort just to raise a B to an A.
•   I will be a better contributor in class, not in the sense that I will be trying to make a good impression on my teachers. I will do whatever I need to do to learn. This may mean placing more demands on teachers to clarify points, making more contributions, and involving myself in discussions with peers.
•   I will be reading more broadly in the area of my major, psychology, not just the articles and books assigned but also in the areas of my interest. I will let my desire to know drive my learning.

She is particularly disappointed in the way she goes about writing the papers assigned in class. She believes that she wastes lots of time and ends up frustrated. And so she picks this topic for this exercise.

> **Declaration of intent**: "I'm going to finally do something about the way I write papers for class."
>
> **Overall aim**: "I'm going to put in place a systematic but flexible process for writing papers for class."

The task of the training group helper is to help Luisa develop and shape a problem-managing goal. Luisa begins by sharing her declaration of intent and overall aim. Here is some of the dialogue that follows:

> **JEREMY**: What bothers you most about the way you have been doing your papers?
> **LUISA**: Everything piles up at the end and as a result the papers are rushed and practically never represent the kind of work I'm capable of.
> **JEREMY**: So the unsystematic approach drives out quality. Let's see what a more systematic approach might look like. What are some of the elements of the process you want to put in place?

LUISA: Well, I've thought about this a bit. For instance, once a paper is assigned, I will start a file on the topic and collect ideas, quotes, and data as I go along.

JEREMY: So you'll start early and have an organized filing system.

LUISA: Yes. I'll start with a single folder and just throw in ideas or quotes from books or parts of lecture notes. Ideas from anywhere or everywhere. I'll sort them out later.

JEREMY: It sounds like the sorting process could become a bottleneck.

LUISA: Hmm. You mean I'll end up with wads of stuff and wonder where to start. Well, there are two things I can do. I'll put only one item to a sheet of paper. One idea. One quote. You know. Then after two or three weeks I'll read through and separate them into different categories and organize them in folders.

JEREMY: The sub-category approach.

LUISA: Right. Later on I can decide which topics are the most important ones. Then I can . . .

Luisa and Jeremy go on in this vein until she has a clear idea of the process that she wants to put in place. When Luisa feels that she has a goal that will drive action, they stop and review and fine-tune the goal in terms of the goal characteristics outlined above. Prepare for your session with your helper by doing the following:

Statement of good intention.

_____

_____

Broad aim.

_____

_____

Now have a goal-setting dialogue based on the criteria for effective goals.

EXERCISE 75: RELATING GOAL CHOICE TO ACTION

Once clients state what they want, they need to move into action in order to get what they want. At this point there are two ways of looking at the relationship of goals to action, one formal and one informal.

Formal Action. Stage III of the helping model—brainstorming action strategies, choosing the best package, and turning them into a plan for accomplishing goals—is the formal approach to action.

Informal Action. Once clients get a fairly clear idea of what they want, there is no reason why they cannot move immediately into action, that is, do _something_ that will move them in the direction of their goals. These are the "little" actions that can precede the formal planning process. Indeed these little actions can help clarify and fine-tune goals. This exercise is about these little actions.

1.   State three personal goals that have emerged from your work in the training group.
2.   For each goal list some "little actions" in which you might be able to engage in order to begin moving quickly in the direction of the goal.
3.   Get feedback from a learning partner on both the goals and the actions.

Example. Review the case of Vanessa discussed above in Exercise 69. While she believes that it is essential for her to develop a better attitude about herself and life in general, she believes that it is not best to try to do this directly. Therefore, she chooses goals which will have a better attitude as a by-product.

- One goal centers around the development of a social life. For the time being, she wants a range of friends rather than an intimate partner or potential spouse. She would especially like a couple of good women friends.
- A second goal revolves around religion in some sense of that term. She would like to have religion-related activities, but not necessarily church-related, as part of her life. Something that deals with the "bigger" questions of life.
- A third goal relates to physical well being. The stress of the divorce has left her exhausted. She wants to build herself up physically and get into good physical shape.

Vanessa looks at the third goal and has this to say: "There are some things I can do immediately without coming up with a formal physical fitness program. First of all exercise. I can begin by walking—to work, to the store, and so on. Second, I can easily cut down a lot on fast food and add some fresh fruit and vegetables to my diet. Let me try these two things and see where it leads."

Now move on to your own personal goals.

First personal goal.

_____

_____

Actions you can take immediately.

_____

_____

_____

_____

_____

Second personal goal.

_____

_____

Actions you can take immediately.

_____

_____

_____

_____

_____

Third personal goal.

_____

_____

Actions you can take immediately.

_____

_____

_____

_____

_____

Chapter 17

STEP II-C: COMMITMENT TO GOALS
"What Are You Willing to Pay for What You Want?"

Many of us choose goals that will help us manage problems and develop opportunities, but we do not explore them from the viewpoint of level of commitment. Just because goals are tied nicely to the original problem situation and initially are espoused by us does not mean that we will follow through. Even goals which seem attractive from a cost/benefit point of view can fall by the wayside. The costs of accomplishing goals might not seem very high until it is time to pay up. Therefore, helping clients take a good look at commitment can raise the probability that they will actually pursue and accomplish these goals. Read Chapter Seventeen in *The Skilled Helper* before doing these exercises.

EXERCISE 76: MANAGING YOUR COMMITMENT TO YOUR GOALS

In this exercise you are asked to review the goals you have chosen to manage some problem situation or pursue some opportunity. The review should focus on commitment. It is not a question of challenging your good will. All of us, at one time or another, make commitments that are not right for us.

1. Review the problem situations and unexploited opportunities you have been examining in this training program together with the goals you have established for yourself.
2. Choose two important goals you have decided to pursue.
3. Read the following questions and answer the ones you see relevant to each goal in order to gauge your level of commitment:

- What is your state of readiness for change in this area at this time?
- To what degree are you choosing this goal freely?
- To what degree are you choosing this goal from among a number of possibilities?
- How appealing is this goal to you?
- Name any ways in which your goal does not appeal to you.
- What's driving you to choose this goal?
- If your goal is in any way being imposed by others, what are you doing to make it your own? What incentives are there besides mere compliance?
- What difficulties are you experiencing in committing yourself to this goal?
- To what degree is it possible that your commitment is not a true commitment?
- What can you do to get rid of the disincentives and overcome the obstacles?
- What can you do to increase your commitment?
- In what ways can the goal be reformulated to make it more appealing?
- To what degree is the timing for pursuing this goal right or wrong?

Evaluation of your commitment to your first goal.

_____

_____

_____

_____

_____

_____

Evaluation of your commitment to your second goal.

_____

_____

_____

_____

_____

_____

4.   With a learning partner, review your principal learnings from answering the above questions about these two goals. Use empathic highlights, probes, and challenges to help one another explore levels of commitment.

5.   Finally, to the degree necessary, reformulate both goals in terms of what you have learned from the dialogue.

Restated goal #1.

_____

_____

_____

_____

_____

Restated goal #2.

_____

_____

_____

_____

_____

EXERCISE 77: REVIEWING BENEFITS VERSUS COSTS IN CHOOSING GOALS

In most choices we make there are both benefits and costs. Commitment to a goal often depends on a favorable cost/benefit ratio. To what extent do the benefits outweigh the costs? The balance-sheet methodology outlined in Chapter Nineteen of *The Skilled Helper* can be useful in determining whether goals being chosen offer "value for money," as it were. Consider the following case.

Case. One January, Helga, a married woman with two children, one a senior in college and one a sophomore, was told that she had an advanced case of cancer. She was also told that a rather rigorous series of chemotherapy treatments might prolong her life, but they would not save her. She desperately wanted to see her daughter graduate from college in June, so she opted for the treatments. Although she found them quite difficult, she buoyed herself up by the desire to be at the graduation. Although in a wheel chair, she was there for the graduation in June. When the doctor suggested that she could now face the inevitable with equanimity, she said: "But, doctor, in only two years my other son will be graduating."

This is a striking example of a woman's deciding that the costs, however high, were outweighed by the benefits. Obviously, this is not always the case. Benefits include incentives and rewards for whatever source, while costs include time and effort spent, resources used, psychological wear and tear, battling negative emotions, and the like. This exercise gives you the opportunity to explore your goals from a cost-benefit perspective. To what degree is it worth the effort? What's the payoff?

1. Divide up into pairs, with one partner acting as client, one as helper.
2. Help your partner review one of the goals of his or her agenda from a cost/benefit perspective. The helper is to use empathic highlights, probing, and challenging to help his or her partner do this. Help your partner identify benefits and costs and do some kind of trade-off analysis such as the balance-sheet technique.
3. Help you partner clearly state the incentives and payoffs that enable her or him to commit to the specified goals.

Example: Carlita, 28-year-old and a third-year doctoral student in clinical psychology, is discussing her commitment to her career goals with Eban, a fellow student. She is an only child and both parents and grandparents have been pressing her to get married and have children. She wants to establish herself in her career first. She has one more year to go in the program plus a year's internship.

> **CARLITA**: If I don't get my degree and begin my career before getting married, I'll never finish. I know others can do this. But I would be feeling all sorts of pressure from family and friends.
> **EBAN**: So, from your perspective, it's almost now or never. But you're already under a lot of pressure. What about the cost-benefit equation?
> **CARLITA**: I'm the first woman to break the mold in my extended family. So I have to expect some costs. I don't like hearing disappointment in the voices of my parents and grandparents. But I'm setting a direction for the rest of my life.
> **EBAN**: You're ready to pay the price. How do you counter the pressures?
> **CARLITA**: First of all, I don't avoid them. I talk to them about how exciting I find my work. I talk to them about how this will benefit both me and whatever family I have in the future. Without saying it in these words, I ask them to be proud of me, not disappointed. I'm still the same daughter and granddaughter that they have always loved. I let them know by my actions that I love them. If I were to give in, that wouldn't be love. In my own way I let them know how *I* need to be loved.
> **EBAN**: What about social life? Any pressures there?

141

CARLITA: I have a couple of close men friends. But they know I'm not going to get into a very serious relationship right now. Eventually, I want someone who takes me as I am.

EBAN: So there are a few bumps there, too.

CARLITA: None that I can't handle. I see myself as responsible and focused. If they want to see that as bullheadedness, well, that's their problem. Furthermore . . .

And the dialogue continues in this vein.

4. After ten minutes of discussion, stop and receive feedback from the client and the person being helped.
5. After the discussion, each member of the training is to get a new partner, change roles, and repeat the process.

EXERCISE 78: COMPETING AGENDAS

When setting goals, clients often fail to see them in the context of the current range of demands on their time, energy, and resources. It is no wonder, then, that some goals do not get accomplished because they are squeezed out by all the other things clients have to do. In other words, clients have to face up to the reality of "competing agendas." Otherwise they end up trying to carry impossible burdens. Therefore, competing agendas need to be reviewed lest they block or dilute goals set during the helping process.

Example. Ivor, 24, is talking to members of his training group about a certain lack of progress in learning both the helping model and the skills that make it work. Now believing he knows what is happening, he says: "I came to this program all fired up to become the best counselor possible. But it hasn't quite happened. I know I'm not lazy. But this past week it struck me. I really enjoy the social life of this group. Lots of us have become friends. So much so that my social life with the group has been doing two things. First, I'm letting it take up a lot of my time. I put off the work I should be doing. Second, I'm reluctant to challenge my friends in both group interactions and in one-to-one practice. This is being unfair to myself and unfair to you. I certainly want to enjoy the social life here, but I don't want it to compete with my original goal."

1. Look into two different goals you have set for yourself over the course of this training program.
2. Review the current commitments of your life both within the training group and outside. Outline the agendas that currently take up a great deal of your time, energy, and interest.
3. Indicate any competing agendas that might prevent you from accomplishing your goal.
4. Outline what needs to be done to manage competing agendas. This could mean dropping some current activities or reducing the scope of the goal.

Competing agendas for first goal.

_____

_____

_____

_____

_____

Ways of managing competing agendas.

_____

_____

_____

_____

_____

Competing agendas for second goal.

_____

_____

_____

_____

Ways of managing competing agendas.

_____

_____

_____

_____

_____

PART FIVE

STAGE III:
HELPING CLIENTS DEVELOP STRATEGIES TO ACCOMPLISH THEIR GOALS

Stage III is an important part of the solution-focused phase of helping. Once clients are helped to establish goals, they often still need help to plan the actions or strategies that will enable them to accomplish their goals. These actions or strategies are solutions with a small s. Chapter 18 proves an introduction to Stage III and discusses strategies in terms of possibilities. Just as Step II-A focused on goals in terms of possibilities for a better future, so Step III-A talks about the many different paths to any given goal. It is about options. Chapter 19, Step III-B, is about helping clients choose the strategies or paths that best fit the needs, style, and resources of the client. It is about choosing the right options. Finally, Step III-C is about taking chosen goals and strategies and turning them into a problem-managing or opportunity-developing plan. Planning, whether formal or informal, plays a key role in problem management and opportunity development.

Stage III has three steps in our usual definition of step. They are all aimed at client action.

Step III-A: Strategies. Help clients develop possible strategies for accomplishing their goals. "What kind of actions will help me get what I need and want?"

Step III-B: Best-Fit Strategies. Help clients choose strategies tailored to their preferences and resources. "What actions are best for me?"

Step III-C: Plans. Help clients formulate actionable plans. "What should my campaign for constructive change look like? What do I need to do first? Second? Third?"

Stage III adds the final pieces to a client's planning a program for constructive change. Stage III deals with the "game plan." However, these three "steps" constitute *planning* for action and should not be confused with action itself. Without action, a program for constructive change is nothing more than a wish list. The formal implementation of plans is discussed in Part VI.

Chapter 18

STEP III-A: ACTION STRATEGIES:
"How Many Ways Are There To Get What I Need And Want?"

There is usually more than one way to accomplish a goal. However, clients often focus on a single strategy or just a few. The task of the counselor in Step III-A is to help clients discover a number of different routes to goal accomplishment. Clients tend to choose a better strategy or set of strategies if they choose from among a number of possibilities. Read Chapter Eighteen before doing the exercises in this section.

LIFE SKILLS

Sometimes people have problems or fail to develop opportunities because they do not have the kinds of **life skills** needed to handle developmental tasks and to invest themselves effectively in the social systems of life. For instance, a young married couple, Roger and Tess, find that they don't have the communication skills needed to talk to each other reasonably about the problems they have been encountering during the first couple of years of marriage. Therefore, in many cases it makes sense to help clients develop the skills they need to achieve their goals. The following is a self-assessment exercise. You are asked to review the kinds of life skills you need both to manage your own problem situations and develop unused opportunities.

EXERCISE 79: ASSESSING CRITICAL SKILLS FOR EFFECTIVE LIVING

This exercise is a checklist designed to help you get in touch with both your resources and possible areas of deficit. Listed below are various groups of skills needed to undertake the tasks of everyday living.

First, rate how important each skill is in your eyes. Use the following scale.

5. For me this skill is **very important**.
4. For me this skill is of **moderate** importance.
3. For me this skill has **average** importance.

2. For me this skill is **rather unimportant**.
1. For me this skill is **not important at all**.

Second, rate yourself on each skill. The rating system is as follows:

5. I have a **very high** level of this skill.
4. I have a **moderately high** level of this skill.
3. From what I can judge, I am about **average** in this skill.
2. I have a **moderate deficit** in this skill.
1. I have a **significant deficit** in this skill.

1. **Body-Related Skills**	Importance	Level
• Knowing how to eat nutritionally.	_____	_____
• Knowing how to control weight.	_____	_____
• Knowing how to keep fit through exercise.	_____	_____
• Knowing how to maintain basic body hygiene.	_____	_____
• Basic grooming skills.	_____	_____
• Knowing what to do when everyday health problems such as colds and minor accidents occur.	_____	_____
• Skills related to sexual expression.	_____	_____
• Athletic skills.	_____	_____
• Aesthetic skills such as dancing.	_____	_____

Other body-related skills:

	Importance	Level
• _____	_____	_____
• _____	_____	_____

2. **Learning-How-to-Learn Skills**	Importance	Level
• Knowing how to read well.	_____	_____
• Knowing how to write clearly.	_____	_____
• Knowing basic mathematics.	_____	_____
• Knowing how to learn and study efficiently.	_____	_____
• Knowing something about the use of computers.	_____	_____
• Using history to understand today's events.	_____	_____
• Being able to use basic statistics.	_____	_____
• Knowing how to use a library.	_____	_____
• Knowing how to find information I need.	_____	_____

Other learning and learning-how-to-learn skills.

	Importance	Level
• _____	_____	_____
• _____	_____	_____

3. Skills Related to Values	**Importance**	**Level**

- Knowing how to clarify my own values to those who have a significant relationship to me.   _____   _____
- Knowing how to identify the values being "pushed" by the social systems to which I belong.   _____   _____
- Knowing how to construct and reconstruct my own set of values.   _____   _____

Other values-related skills:

- _____   _____   _____

- _____   _____   _____

4. Self-Management Skills	**Importance**	**Level**

- Knowing and understanding myself.   _____   _____
- Knowing how to plan and set realistic goals.   _____   _____
- Problem-solving and opportunity-development skills.   _____   _____
- Decision-making skills.   _____   _____
- Knowing how to make incentives, punishments, and rewards work for me rather than against me.   _____   _____
- Managing my emotions.   _____   _____
- The ability to delay gratification.   _____   _____
- Assertiveness: getting my needs met while respecting the legitimate needs of others.   _____   _____

Other self-management skills:

- _____   _____   _____

- _____   _____   _____

5. Communication Skills	**Importance**	**Level**

- The ability to listen to others actively.   _____   _____
- The ability to communicate understanding to others (empathy).   _____   _____
- The ability to challenge others reasonably.   _____   _____
- The ability to provide useful information to others.   _____   _____
- The ability to explore with another person what is happening in my relationship with him or her.   _____   _____
- The ability to speak before a group.   _____   _____

Other communication skills:

- _____   _____   _____

- _____   _____   _____

Now that you have done a brief assessment of some life skills, indicate which skills, if improved, would help you manage your concerns, problems, or soft spots better and develop opportunities more effectively. When you name a skill, indicate why such a skill is important to you and what you might do to develop it.

_____

_____

_____

_____

_____

_____

_____

What kinds of life skills do you think you need to become better at, not just to handle your own problems more effectively, but to be an effective counselor?

_____

_____

_____

_____

End by discussing with a learning partner what you have discovered about yourself through this self-assessment.

EXERCISE 80: BRAINSTORMING ACTION STRATEGIES FOR YOUR OWN GOALS

Brainstorming is a technique you can use to help yourself and your clients move beyond constricted to imaginative or inventive thinking. Recall the rules of brainstorming:

- Encourage quantity. Deal with the quality of suggestions later.
- Do not criticize any suggestion. Merely record it.
- Combine suggestions to make new ones.
- Encourage wild possibilities—"One way to keep to my diet and lose weight is to have my mouth sewn up."
- When you feel you have said all you can say, put the list aside and come back to it later and add fresh ideas.

Example: Ira, a retired lawyer in training to be a counselor, is in a high-risk category for a heart attack: some of his relatives have died relatively early in life from heart attacks, he is overweight, he exercises very little, he is under a great deal of pressure in his job, and he smokes over a pack of cigarettes a day.

One of his goals is to stop smoking within a month. With the help of a nurse practitioner friend, he comes up with the following list of strategies:

- just stop cold turkey.
- shame myself into it, "How can I be a helper if I engage in self-destructive practices such as smoking?"
- cut down, one less per day until zero is reached.
- use the new nicotine-based aids, such as patches and gum.
- look at movies of people dying with lung cancer.
- pray for help from God to quit.
- use those progressive filters which gradually squeeze all taste from cigarettes.
- switch to a brand that doesn't taste good.
- switch to a brand that is so heavy in tars and nicotine that even I see it as too much.
- smoke constantly until I can't stand it any more.
- let people know that I'm quitting.
- put an ad in the paper in which I commit myself to stopping.
- send a dollar for each cigarette smoked to a cause I don't believe in, for instance, the "other" political party.
- get hypnotized; through a variety of post-hypnotic suggestions have the craving for smoking lessened.
- pair smoking with painful electric shocks.
- take a pledge before my minister to stop smoking.
- join a group for professionals who want to stop smoking.
- visit the hospital and talk to people dying of lung cancer.
- if I buy cigarettes and have one or two, throw the rest away as soon as I come to my senses.
- hire someone to follow me around and make fun of me whenever I have a cigarette.
- have my hands put in casts so I can't hold a cigarette.
- don't allow myself to watch television on the days in which I have even one cigarette.
- reward myself with a week-end fishing trip once I have not smoked for two weeks.
- avoid friends who smoke.
- have a ceremony in which I ritually burn whatever cigarettes I have and commit myself to living without them.
- suck on hard candy made with one of the non-sugar sweeteners instead of smoking.
- give myself points each time I want to smoke a cigarette and don't; when I have saved up a number of points reward myself with some kind of "luxury."

Note that Ira includes a number of wild possibilities in his brainstorming session. Now follow these steps.

1. On separate paper, brainstorm ways of achieving each of the goals you list below. Add wilder possibilities at the end.
2. After you have finished your list, take one of the goals and the brainstorming list, sit down with a learning partner, and through dialogue expand your list. Follow the rules of brainstorming.
3. Switch roles. Through empathic highlights, probing, summarizing, and challenge, help your partner expand his or her list.
4. Keep both lists of brainstormed strategies. You will use them in a later exercise.

Goal # 1. _____

_____

Goal # 2. _____

_____

EXERCISE 81: ACTION STRATEGIES: PUTTING YOURSELF IN THE CLIENT'S SHOES

Here are a number of cases in which the client has formulated a goal and needs help in determining how to accomplish it. You are asked to put yourself in the client's shoes and brainstorm action strategies that you yourself might think of using were you that particular client.

Example: Renalda, 53, has been a very active person—career-wise, physically, socially, and intellectually. In fact, she has always prided herself on the balance she has been able to maintain in her life. However, an auto accident that was not her fault has left her a paraplegic. With the help of a counselor she has begun to manage the depression that almost inevitably follows such a tragedy. In the process of re-directing her life, she has set some goals. Since her job and her recreational activities involved a great deal of physical activity, a great deal of re-direction is called for.

One of her goals is to write a book called "The Book of Hope" about ordinary people who have creatively re-set their lives after some kind of tragedy. The book has two purposes. Since it would be partly autobiographical, it would be a kind of chronicle of her own re-direction efforts. This will help her commit herself to some of the grueling rehabilitation work that is in store for her. Second, since the book would also be about others struggling with their own tragedies, these people will be models for her. Renalda has never published anything, so the "how" is more difficult. For her, writing the book is more important than publishing it. Therefore, the anxiety of finding a publisher is not part of the "how."

Hobart is a graduate student in a clinical psychology training program. He puts himself in Renalda's shoes and asks himself, "What would I have to do to get a book written?" Here are some of his brainstormed possibilities:

- Get a book on writing and learn the basics.
- Start writing short bits on my own experience, anything that comes to mind.
- Read books written by those who conquered some kind of tragedy.
- Talk to the authors of these books.
- Find out what the pitfalls of writing are like, for instance, writer's block.
- Get a ghost writer who can translate my ideas into words.
- Write a number of very short, to-the-point pamphlets, then turn them into a book.
- Learn how to use a word-processing program both as part of my physical rehabilitation program and as a way of jotting down and playing with ideas.
- Do rough drafts of topics that interest me and let someone else put them into shape.
- Record discussions about my own experiences with the counselor, the rehabilitation professionals, and friends and then have these transcribed for editing.
- Interview people who have turned tragedies like mine around.
- Interview professionals and the relatives and friends of people involved in personal tragedies. Record their points of view.
- Through discussion with friends get a clear idea of what this book will be about.
- Find some way of making it a bit different from similar books. What could I do that would give such a book a special slant?

Now do the same kind of work for each of the following cases.

Case # 1. Emma, 27, has been a heroin addict. Her life has revolved around her habit. Her employer, an ad agency has told her, that, despite her creativity, her erratic performance at work will no longer be tolerated. She opts for a treatment—known as ultrarapid opiate detoxification—in which the drug naltrexone replaces heroin in the addict's opium receptors within six hours or so. Treatment with naltrexone is to continue for six months. Now that she is free of opiates, the challenge is to build a very different lifestyle—one that will focus her talents and energies. She is in opportunity-development mode. This includes a new social life since, currently, she has no social life to speak of. One of her goals is reconciliation with her family, including her parents, one sister, and two brothers together with their families. She has seen very little of them over the past three years apart from attending two weddings. She wants to start the work on reconciliation while she is still feeling well. She realizes that reconciliation is a two-way street and that she cannot set goals for others.

1. **Goal.** If you were Emma, what would "reconciliation with my family" look like? If accomplished what would this goal look like? What would be in place that is not now in place? How is the goal modified by the fact that reconciliation is a two-way street? Formulate a goal that has the characteristics of a viable goal outlined earlier.

_____

_____

_____

_____

_____

2. **Brainstorming Strategies.** What are some of the things you might do to achieve the kind of reconciliation with your family that you have outlined above? Do the brainstorming on a separate sheet of paper. Remember to include some "wild" possibilities.
3. After you have developed your list, first share your goal with a learning partner. Give each other feedback on the quality of the goal. Note especially how you and your partner interpret "reconciliation."
4. Next share the lists of brainstormed possibilities. Working together, add several more possibilities to the combined lists. Keep your list for use in an exercise in Step III-B.
5. Discuss with your partner what you have learned from this exercise.

Case # 2. A priest was wrongfully accused of molesting a boy in his parish. Working with a counselor, he set three goals. His "now" goal was to maintain his equilibrium under stress. Since, at the time other priests had been accused and convicted of pedophilia, he knew that in the eyes of many he would be seen to be guilty until proved innocent. With the help of the counselor and some close friends, both lay and clerical, he kept his head above water. A near-term goal was to win the case in court. He also accomplished this goal. He was acquitted and all charges were dropped. After the trial the bishop wanted to send him to a different parish "to start fresh." He had been removed from his pastorate and was living at the seminary. But he wanted to return to the same parish and re-establish his relationship with his parishioners. After all, he had done nothing wrong. The bishop agreed to reinstate him in his parish.

1. **Goal.** If you were this man, what would "re-establishing my relationship with reconciliation with my parishioners" look like? What would be in place that is not now in place? Formulate a goal that has the characteristics of a viable goal outlined earlier.

_____

_____

_____

_____

_____

2.   **Brainstorming Strategies.** What are some of the things you might do to re-establish your relationship with your parishioners, especially in view of the fact that some might still see you as tainted by the whole affair? Do the brainstorming on a separate sheet of paper. Include some "wild" possibilities.

Now complete steps 3-5 as above.

EXERCISE 82: HELPING OTHERS BRAINSTORM STRATEGIES FOR ACTION

As a counselor, you can help your fellow trainees stimulate their imaginations to come up with creative ways of achieving their goals. In this exercise, use probes and challenges based on questions such as the following:

- **How:** How can you get where you'd like to go? How many different ways are there to accomplish what you want to accomplish?
- **Who:** Who can help you achieve your goal? What people can serve as resources for the accomplishment of this goal?
- **What:** What resources both inside yourself and outside can help you accomplish your goals?
- **Where:** What places can help you achieve your goal?
- **When:** What times or what kind of timing can help you achieve your goal? What times might be better than others?

Example: What follows are bits and pieces of a counseling session in which Angie, the counselor trainee, is helping Meredith, the client, develop strategies to accomplish one of his goals. One of Meredith's problems is that he procrastinates a great deal. He feels that he needs to manage this problem in his own life if he is to help future clients move from inertia to action. Therefore, his "good intention" is to reduce the amount of procrastination in his life. In exploring his problem, he realizes that he puts off many of the assignments he receives in class. The result is that he is overloaded at the end of the semester, experiences a great deal of stress, does many of the tasks poorly, and receives lower grades than he is capable of. While his overall goal is to reduce the total amount of procrastination in his life, his immediate goal is to be up-to-date every week in all assignments for the counseling course. He also wants to finish the major paper for the course one full week before it is due. He chooses this course as his target because he finds it the most interesting and has many incentives for doing the work on time. He presents his list of the strategies he has brainstormed on his own to Angie. After discussing this list, their further conversations sound something like this:

      ANGIE: You said that you waste a lot of time. Tell me more about that.
      MEREDITH: Well, I go to the library a lot to study. And I go with the best intentions. But, I meet friends; we kid around, and time slips away. I guess the library is not the best place to study.
      ANGIE: Does that suggest another strategy?
      MEREDITH: Yeah, study someplace where none of my friends is around. But, then . . . I might not . . .

ANGIE (interrupting): We'll evaluate this later. Right now let's just add it to the list.

* * * * *

ANGIE: I know it's your job to manage your own problems, but I assume that you could get help from others and still stay in charge of yourself. Could anyone help you achieve your goal?

MEREDITH: I've been thinking about that. I have one friend . . . we make a bit of fun of him because he makes sure he gets everything done on time. He's not the smartest one of our group, but he gets good grades because he knows how to study. I'd like to pair up with him in some way, maybe even anticipate deadlines the way he does.

* * * * *

ANGIE: Your strategy list sounds a bit tame. Maybe it sounds wild to you because you're trying to change what you do.

MEREDITH: I guess I could get wilder. Hmm . . . I could make a contract with my counseling prof to get the written assignments in early! That would be wild for me.

Note here that Angie uses probes based on the how, who, what, when, and where probes outlined above. She also uses the "wilder possibilities" probe.

1. Divide up into groups of three: client, helper, and observer.
2. Decide in which order you will play these roles.
3. The client will briefly summarize a concern or problem and a specific goal, which, if accomplished, will help him, or her manage the problem or develop the unused opportunity more fully. Make sure that the client states the goal in such a way that it fulfills the criteria for a viable goal.
4. Give the client five minutes to write down as many possible ways of accomplishing the goal as he or she can think of.
5. Then help the client expand the list. Use probes and challenges based on the questions listed above.
6. Encourage the client to follow the rules of brainstorming. For instance, do not let him or her criticize the strategies as he or she brainstorms.
7. At the end of the session, stop and receive feedback from your client and the observer as to the helpfulness of your probes and challenges.
8. Switch roles and repeat the process until each has played all three roles.

Chapter 19

STEP III-B: BEST-FIT STRATEGIES--
What Actions Are Best For You?

The principle is simple. Strategies for action chosen from a large pool of strategies tend to be more effective than those chosen from a small pool. However, if brainstorming is successful, clients are sometimes left with more possibilities than they can handle. Therefore, once clients have been helped to brainstorm a range of strategies, they might also need help in choosing the most useful. These exercises are designed to help you help clients choose "best-fit" strategies, that is, strategies that best fit the resources, style, circumstances, and motivation level of clients.

EXERCISE 83: A PRELIMINARY SCAN OF BEST-FIT STRATEGIES FOR YOURSELF

You do not necessarily need sophisticated methodologies to come up with a package of strategies that will help you accomplish a goal. In this exercise you are asked to use your common sense to make a "first cut" on the strategies you brainstormed for yourself in Exercise 79 (Chapter 18).

1.   Review the strategies you brainstormed for each of the goals in that exercise.
2.   Star the strategies that make most sense to you. Just use common-sense judgment. Use the following example as a guideline.

Example: Let's return to the case of Ira in Exercise 79. Remember that he is the counselor trainee who wants to stop smoking. Here are the strategies he brainstormed. The ones he chooses in a preliminary common-sense scan are marked with a check mark (✓) instead of a bullet (●).

✓   just stop cold turkey.
✓   shame myself into it, "How can I be a helper if I engage in self-destructive practices such as smoking?"
●   cut down, one less per day until zero is reached.
✓   use the new nicotine-based aids, like gum, advertised on TV.
●   look at movies of people dying with lung cancer.
✓   pray for help from God to quit.
●   use those progressive filters which gradually squeeze all taste from cigarettes.
●   switch to a brand that doesn't taste good.
●   switch to a brand that is so heavy in tars and nicotine that even I see it as too much.
●   smoke constantly until I can't stand it any more.
●   let people know that I'm quitting.
●   put an ad in the paper in which I commit myself to stopping.
●   send a dollar for each cigarette smoked to a cause I don't believe in, for instance, the "other" political party.
●   get hypnotized; through a variety of post-hypnotic suggestions have the craving for smoking lessened.
●   pair smoking with painful electric shocks.
●   take a pledge before my minister to stop smoking.
●   join a group for professionals who want to stop smoking.
●   visit the hospital and talk to people dying of lung cancer.
●   if I buy cigarettes and have one or two, throw the rest away as soon as I come to my senses.
●   hire someone to follow me around and make fun of me whenever I have a cigarette.

- have my hands put in casts so I can't hold a cigarette.
- don't allow myself to watch television on the days in which I have even one cigarette.
- ✓ reward myself with a week-end fishing trip once I have not smoked for two weeks.
- have a ceremony in which I ritually burn whatever cigarettes I have and commit myself to living without them.
- suck on hard candy made with one of the non-sugar sweeteners instead of smoking.
- ✓ give myself points each time I want to smoke a cigarette and refrain from doing so; when I have saved up a number of points reward myself with some kind of "luxury."

3. Share with a learning partner the strategies you have starred for yourself and the reasons for choosing them. What makes each of them "best-fit" for you?

EXERCISE 84: BEST-FIT STRATEGIES: PUTTING YOURSELF IN THE CLIENT'S SHOES

In this exercise you are asked to put yourself in clients' shoes as they struggle to choose the strategies that will best enable them to accomplish their goals.

1. Review the case of Renalda in Exercise 80 (Chapter 18).
2. Review the strategies that were brainstormed by Hobart, the clinical psychology trainee who put himself in Renalda's shoes:

- Get a book on writing and learn the basics.
- Start writing short bits on my own experience, anything that comes to mind.
- Read books written by those who conquered some kind of tragedy.
- Talk to the authors of these books.
- Find out what the pitfalls of writing are like writer's block.
- Get a ghost writer who can translate my ideas into words.
- Write a number of very short, to-the-point pamphlets, then turn them into a book.
- Learn how to use a word-processing program both as part of my physical rehabilitation program and as a way of jotting down and playing with ideas.
- Do rough drafts of topics that interest me and let someone else put them into shape.
- Record discussions about my own experiences with the counselor, the rehabilitation professionals, and friends and then have these transcribed for editing.
- Interview people who have turned tragedies like mine around.
- Interview professionals and the relatives and friends of people involved in personal tragedies. Record their points of view.
- Through discussion with friends get a clear idea of what this book will be about.
- Find some way of making it a bit different from similar books. What could I do that would give such a book a special slant?

3. If you have further strategies, add them to the list now.
4. Using your common sense, star the strategies you believe belong in the best-fit category.
5. Jot down the reasons for your choices. What makes them best-fit?
6. Share your choices and reasons with a learning partner and give each other feedback.

EXERCISE 85: USING CRITERIA TO CHOOSE BEST-FIT STRATEGIES

Just as there are criteria for crafting goals (Step II-B), so are there criteria for choosing best-fit strategies. The following questions can be asked especially when the client is having difficulty choosing from among a number of possibilities. These criteria complement rather than take the place of common sense.

- **Clarity.** Is the strategy clear?
- **Relevance.** Is it relevant to my problem situation and goal?
- **Realism.** Is it realistic? Can I do it?
- **Appeal.** Does it appeal to me?
- **Values.** Is it consistent with my values?
- **Efficacy.** Is it effective enough? Does it have bite? Will it get me there?

Example: Ira, the counselor trainee who wanted to quit smoking, considered the following possibility on his list: "Cut down gradually, that is, every other day eliminate one cigarette from the 30 I smoke daily. In two months, I would be free."

- **Clarity:** "This strategy is very clear; I can actually see the number diminishing. It gets a 6 or 7 for clarity."
- **Relevance:** "It leads inevitably to the elimination of my smoking habit, but only if I stick with it."
- **Realism:** "I could probably bring this off. It would be like a game; that would keep me at it. But maybe too much like a game. Also I know others who have done it."
- **Appeal:** "Although I like the idea of easing into it, I'm quitting because I now am personally convinced that smoking is very dangerous for me. I should stop at once."
- **Values:** "There is something in me that says that I should be able to quit cold turkey. That has more moral appeal to me. For me there's something phony about gradually cutting down."
- **Effectiveness:** "The more I draw this action program out, the more likely am I to give it up. There are too many pitfalls spread out over a two-month period."

In summary, Ira says, "I now see that only strategies related to stopping cold turkey have bite. In fact, stopping is not the hard thing. Not taking smoking up again in the face of temptation, that's the real problem." The above criteria not only helped Ira eliminate all strategies related to a gradual reduction in smoking, but it helped him redefine his goal to "stopping and staying stopped." Sustainability is the real issue. He also noted that the brainstormed strategies he preferred referred not just to stopping but to sustainability. For instance, in the short term, the nicotine-flavored gum would reduce the craving once he has stopped.

1.   Read the following case, put yourself in this woman's shoes, and, like Ira, use the criteria for determining the viability of the strategy she proposes.

Case: A young woman has been having disagreements with a male friend. Since he is not the kind of person she wants to marry, her goal is to establish a relationship with him that is less intimate, for instance, one without sexual relations. She knows that she can be friends with him but is not sure if he can be just a friend with her. She would rather not lose him as a friend. She also knows that he sees other women. She uses the above criteria to evaluate the following strategy: "I'll call a moratorium on our relationship. I'll tell him that I don't want to see him for four months. After that we will be in a better position to re-establish a different kind of relationship, if that's what we want."

2.   Once you have done your analysis, share your findings with a learning partner. See what the two of you can learn from your differences.

DO NOT READ # 3 BEFORE DOING STEPS 1 AND 2 OF THIS EXERCISE!

3.   Now review the reasoning that the woman herself went through and her decision. Here, then, is her analysis:

- **Clarity:** "A moratorium is quite clear; it would mean stopping all communication for four months. It would be as if one of us were in Australia for four months. But no phone calls."
- **Relevance:** "Since my goal is moving into a different kind of relationship with him, stepping back to let old ties and behaviors die a bit is essential. A moratorium is not the same as ending a relationship. It leaves the door open. But it does indicate that cutting the relationship off completely could ultimately be the best course."
- **Realism:** "I can stop seeing him. I think I have the assertiveness to tell him exactly what I want and stick to my decision. Obviously I don't know how realistic he will think it is. He might see it as an easy way for me to brush him off. He might get angry and tell me to forget about it."
- **Appeal:** "The moratorium appeals to me. It will be a relief not having to manage my relationship with him for a while."
- **Values:** "There is something unilateral about this decision and I prefer to make decisions that affect another person collaboratively. On the other hand, I do not want to string him along, keeping his hopes for a deeper relationship and even marriage alive."
- **Effectiveness:** "When--and if, because it depends on him, too--we start seeing each other again, it will be much easier to determine whether any kind of meaningful relationship is possible. A moratorium will help determine things one way or another." Based on her analysis, she decides to propose the moratorium to her friend.

4.   Discuss her reasoning with your learning partner. In what ways does it differ from your own? If you were her helper, in what ways would you challenge her reasoning and her decision?
5.   Use these criteria to review the strategies you chose for yourself in Exercise 82.

EXERCISE 86: CHOOSING BEST-FIT STRATEGIES: THE BALANCE-SHEET METHOD

The balance sheet is another tool you can use to evaluate different program possibilities or courses of action. It is especially useful when the problem situation is serious and you are having difficulty rating different courses of action. Review the balance-sheet format in Chapter Nineteen of *The Skilled Helper*.

Example: Rev. Alex M. has gone through several agonizing months re-evaluating his career in the ministry. He finally decides that he wants to leave the ministry and get an ordinary job. His decision, though painful in coming, leaves him with a great deal of peace. He now wonders just how to go about this. One possibility, now that he has made his decision, is to leave immediately. However, since this is a serious choice, he wants to look at it from all angles. He uses the balance sheet to evaluate the strategy of leaving his position at his present church immediately. We will not present his entire analysis (indeed, each bit of the balance sheet need not be used). Here are some of his key findings:

Benefits for me: Now that I've made my decision to leave, it will be a relief to get away. I want to get away as quickly as possible.

- **Acceptability:** I have a right to think of my personal needs. I've spent years putting the needs of others and of the institution ahead of my own. I 'm not saying that I regret this. Rather, this is now my season.

- **Unacceptability:** Leaving right away seems somewhat impulsive to me, meeting my own needs to be rid of a burden.

Costs for me: I don't have a job and I have practically no savings. I'll be in financial crisis.

> **Acceptability:** My frustration is so high that I'm willing to take some financial chances. Besides, I'm well educated and the job market is good.
> **Unacceptability:** I will have to forego some of the little luxuries of life for a while, but that's not really unacceptable.

Benefits for significant others: The associate minister of the parish would finally be out from under the burden of these last months. I have been hard to live with. My parents will actually feel better because they know I've been pretty unhappy.

- **Acceptability:** My best bet is that the associate minister will be so relieved that he will not mind the extra work. Anyway, he's much better than I at getting people involved in the work of the congregation.
- **Unacceptability**: I can't think of any particular downside here.

Costs to significant social settings: This is the hard part. Many of the things I do in this church are not part of programs that have been embedded in the structure of the church. They depend on me personally. If I leave immediately, many of these programs will falter and perhaps die because I have failed to develop leaders from among the members of the congregation. There will be no transition period. The congregation can't count on the associate minister taking over, since he and I have not worked that closely on any of the programs in question.

- **Acceptability**: The members of the congregation need to become more self-sufficient. They should work for what they get instead of counting so heavily on their ministers.
- **Unacceptability:** Since I have not worked at developing lay leaders, I feel some responsibility for doing something to see to it that the programs do not die. Some of my deeper feelings say that it isn't fair to pick up and run.

Alex goes on to use this process to help him make a decision. He finally decides to stay an extra three months and spend time with potential leaders within the congregation. He will tell them his intentions and then help them take ownership of essential programs.

1.   Choose a personal goal for which you have brainstormed strategies.
2.   Choose a major strategy or course of action you would like to explore much more fully. If this exercise is to be meaningful, the problem area, the goal, and the strategy or course of action in question must have a good deal of substance to them. Using the balance-sheet methodology for a trivial issue would be a waste of time.
3.   Identify the "significant others" and the "significant social settings" that would be affected by your choice.
4.   Explore the possible course of action by using as much of the balance sheet as is necessary to help you make a sound decision.
5.   Share your balance sheet with a learning partner. Through dialogue, give one another feedback and help one another improve the choices.

Chapter 20

STEP III-C:
HELPING CLIENTS MAKE PLANS:
"What Kind Of Plan Will Help Me Get
What I Need And Want?"

A plan is a step-by-step procedure for accomplishing each goal of a change agenda. The strategies chosen in Step III-B often need to be organized and translated into a step-by-step plan. The plan should include only the kind of detail needed to drive action. Overly-detailed plans usually fall by the wayside. Clients are more likely to act if they know what they are going to do first, what second, what third, and so forth. Realistic time frames for each of the steps are also essential. The plan imposes the discipline clients need to get things done. To prepare for these exercises, read Chapter Twenty in the text.

EXERCISE 87: DEVELOPING YOUR OWN ACTION PLAN

In this exercise you are asked to establish a workable plan to accomplish one of the goals you have set for yourself.

1.   Choose a problem situation or undeveloped opportunity that you have been working on, that is, one that you have explored and for which you have developed goals and action strategies.
2.   Indicate the goal you want to accomplish.
3.   Write a simple plan that embodies the principles outlined in Chapter Twenty.

Example. Emma, the 27-year-old heroin addict who has just undergone the ultra-rapid opiate detoxification process and is now opiate-free (see Chapter 18), wants to establish a social life that will include family and former friends. While on heroin, she either abandoned family and friends or they abandoned her. She has explored some ways of going about this with her counselor and settles on the following plan:

•   She will contact a couple of friends and a couple of members of the family who she thinks might be most receptive to her overtures. She will first write them and tell them what has happened and what she wants to do. She will follow up with a phone call.
•   Since she realizes that she might experience some rejection, she tells the members of her counseling group that she will need their support. Their support both face-to-face and by phone is an important part of her plan.
•   When she is sure of the help of both a friend and of a family member, she will arrange to meet them separately face-to-face. She will use these encounters to get the "lie of the land" and to brainstorm further possibilities.
•   She will use these initial reconciliations as a basis for further contacts. Hopefully her early contacts will also act as informal messengers to others within her family and among her former friends.

This is far as she wants to go for the moment in the planning process. She will begin to develop further plans only after a couple of successful contacts.

4.   Before doing your own plan, get together with a learning partner and critique the plan that Emma has developed. What are its strong points? What are its deficiencies and how might they be remedied? What might you add to the plan?

The first goal you want to accomplish.

_____

_____

On a separate sheet of paper outline the major steps of your plan.

The second goal you want to accomplish.

_____

_____

On a separate sheet of paper outline the major steps of your plan.

EXERCISE 88: SHARING AND SHAPING PLANS

In this exercise you are asked to share and improve the plan you have developed and to help a learning partner do the same.

1.   Share your written plans with a learning partner. Each of you should read the other's before you get together.
2.   In a face-to-face session, explore and provide feedback to each other on the quality of each plan. How well shaped is it? To what degree is it simple without being simplistic? How concrete is it? What kind of compromise does it reach between too much and too little detail? How good is the fit between the plan and the person? Use empathic highlights, probes, and challenge to help the other shape his or her plan.
3.   In the light of the feedback and dialogue, indicate what changes you would make in the plan.

The changes I would make in my first plan.

_____

_____

_____

_____

_____

The changes I would make in my second plan.

_____

_____

_____

_____

_____

EXERCISE 89: FORMULATING PLANS FOR THE MAJOR STEPS OF A COMPLEX GOAL

If a goal is complex, for instance, changing careers or doing something about a deteriorating marriage, the plan to achieve it will often have a number of major steps. In this case a divide-and-conquer strategy is useful. That is, the complex goal—say, the improvement of the marriage—can be divided up into a number of subgoals. For instance, a more equitable division of household chores might be such a subgoal. It is one goal in the total "package" of goals that will constitute the improved marriage.

In this exercise you are asked to spell out the action steps for two of the subgoals leading up to the accomplishment of some complex goal of your own. Consider this example.

Example: Lynette, a clinical psychology student in a counselor training program, has discovered that she comes across as quite manipulative both to her instructors and to her classmates. She believes that she has developed the style as a response to the less than enthusiastic reception she received whenever she encroached on "male" territory, whether at home, school, or in the workplace. Her current style in working with others is to make the decisions herself while letting the other party think that he or she is having a say. Then, if she does not get her way, she moves to a more overtly domineering style. However, she realizes that this style is contrary to the values she wants to permeate her relationships, including, of course, her relationships with her future clients.

Since this is the first time that she has received such feedback, one of the sub-goals she sets for herself is to get a very concrete understanding of her present style and to develop a better picture of a preferred style. This would be a major step in changing her overall manipulative style. Here are the steps in her plan to implement this style change:

- To elicit the cooperation of her instructor and the members of her training group. She wants to use the training experience as a lab for personal change.
- To spend three weeks interacting as she usually does. That is, for three weeks she will follow her instincts. She knows that she can't pull this off perfectly, because she is now observing herself more critically as are the members of her group.
- To identify "live" instances when her domineering or manipulative style takes over. She can either catch herself or get immediate feedback from others when these instances take place. She will also keep track of instances of manipulative behavior in her everyday life.
- To identify "on the spot," but briefly, how the interchange with another person or the group might have gone better. This will provide her with a practical inventory of possibilities for a change in style.
- At the end of three weeks to draw up a portrait of the dysfunctional style and to pull together a portrait of the preferred style from the possibilities developed within the group.

Move on to the following tasks:

1.   With a learning partner evaluate the steps she has laid out. If she were to ask you for your help, what feedback would you give her about her plan? What changes might you suggest?
2.   Now do the same for two of the major steps of a plan you have developed to achieve some complex goal related to becoming a more effective helper.

A key goal of mine.

_____

_____

One major step in the plan to accomplish this goal.

_____

_____

_____

_____

A brief description of the steps I would take to achieve this subgoal.

_____

_____

_____

_____

_____

_____

_____

A second major step in the plan to accomplish this goal.

_____

_____

_____

A brief description of the steps I would take to achieve this second subgoal.

_____

_____

_____

_____

_____

_____

EXERCISE 90: DEVELOPING THE RESOURCES TO IMPLEMENT YOUR PLANS

Plans can be venturesome, but they must also be realistic. Most plans call for resources of one kind or another. In this exercise you are asked to review some of the goals you have established for yourself in previous exercises and the plans you have been formulating to implement these goals with a view to asking yourself, "What kind of resources do I need to develop to implement these plans?" For instance, you may lack the kinds of skills needed to implement a program. If this is the case, the required skills constitute the resources you need.

1. **Indicate a goal** you would like to implement in order to manage some problem situation or develop some opportunity. Consider the following example. Mark is trying to manage his physical well-being better. He has headaches that disrupt his life. "Frequency of headaches reduced" is one of his goals, a major step toward getting into better physical shape. "The severity of headaches reduced" is another.
2. **Outline a plan** to achieve this goal. The plan may call for resources you may not have or may not have as fully as you would like. Consider Mark once more. Relaxing both physically and psychologically at times of stress and especially when he feels the "aura" that indicates a headache is on its way is one strategy for achieving his goal. Discovering and using the latest drugs to help him control his kind of headache is another part of his plan.
3. **Indicate the resources** you need to develop to implement the plan. For instance, Mark needs the skills associated with relaxing. Furthermore, since he allows himself to become the victim of stressful thoughts, he also needs some kind of thought-control skills. He does not possess either set of skills. Finally, he needs a doctor with whom he can discuss the kind of headaches he gets and possible drugs available for helping control them.
4. **Summarize a plan** that would enable you to develop some of the resources you need to move forward. In one program offered through the school, Mark learns the skills of systematic relaxation and skills related to controlling self-defeating thoughts. The Center for Student Services refers him to a doctor who specializes in headaches.

Problem situation #1

Your problem-managing or opportunity-developing goal.

_____

_____

_____

Skills or other resources you need to accomplish this goal.

_____

_____

Summarize a plan that can help you develop or get the resources you need.

_____

_____

_____

_____

_____

_____

Problem situation #2

Your problem-managing or opportunity-developing goal.

_____

_____

_____

Skills or other resources you need to accomplish this goal.

_____

_____

_____

Summarize a plan that can help you develop or get the resources you need.

_____

_____

_____

_____

_____

_____

5. **Share** your resource-development plans with a learning partner and provide feedback to each other.

PART SIX

THE ACTION ARROW: MAKING IT ALL HAPPEN

The Action Arrow of the helping model indicates the difference
between planning and action. Stages I, II, III and their nine steps
all revolve around planning for change, not change itself.
However, the need to incorporate action into planning and
planning into action has been emphasized throughout the book.
That is, the "little actions" needed to get the change process
moving right from the start have been noted and illustrated. We
now take a more formal look at results-producing action. It is
important to identify both the obstacles to action and ways to
overcome these obstacles.

Chapter 21

"How Do I Make It All Happen?"
HELPING CLIENTS GET WHAT THEY WANT AND NEED

Stages I, II, and III are all planning stages. They are all just a lot of blah, blah, blah if the client does not
engage in and persist in the kind of activity that accomplishes goals. Some clients, once they have a clear
idea of what to do to handle a problem situation—whether or not they have a formal plan—go ahead and
do it. They need little or nothing in terms of further support and challenge from their helpers. They either

find the resources they need within themselves or get support and challenge from the significant others in the social settings of their lives. At the other end of the spectrum are clients who choose goals and come up with strategies for implementing them but who are, for whatever reason, stymied when it comes to action. Most clients fall between these two extremes. There are many reasons why clients fail to act on their own behalf:

- helpers who do not have an action mentality, who help clients discuss problems, but who do not help them act.
- client inertia, that is, clients fail to get started because change is hard work and there are obstacles in the way of constructive change.
- client entropy, that is, the tendency of clients to slow down and give up once they have started.

Often a failure to see serious down-the-road obstacles to action contributes to both inertia and entropy. Chapter Twenty-One, which deals with these and other implementation issues, should be read before doing the exercises below.

Since the skilled-helper model is action-oriented, it is important for you to understand the place of "getting things done" in your own life. If you are to be a catalyst for problem-managing and opportunity-developing action on the part of clients, then reviewing your own track record in this regard is important. Unfortunately, helping often suffers from too much talking and not enough doing. Research shows that helpers are sometimes more interested in helping clients develop new insights than encouraging them to act on them. Inertia and procrastination plague most of us. The exercises in this section are designed to help you explore your own orientation toward action so that you may become a more effective stimulus to action for your clients.

EXERCISE 91: EXPLORING YOUR OWN TENDENCY TO PROCRASTINATE

Procrastination, putting things off, is part of the human condition. We all do it, at least sometimes. In fact, we can become quite inventive in procrastination. Some procrastination is chronic— "I never seem to get around to doing papers for my courses until the last minute." But procrastination can also be incident or project specific—"I know I promised to go see my ailing aunt who loves me dearly, but I haven't gotten around to doing it." In this exercise you are asked to explore either chronic or incident-specific procrastination in your own life.

An example of chronic procrastination: Dahlia, 52, whose children, including one autistic son, are now grown, has returned to school in order to become a counselor. She has this to say about her tendency to procrastinate. "My husband is in business for himself. I take care of a lot of the routine correspondence for the business and our household. Often I let it pile up. The more it piles up the more I hate to face it. On occasion, an important business letter gets lost in the shuffle. This annoys my husband a great deal. Then with a great deal of flurry, I do it all and for a while keep current. But then I slide back into my old ways. I also notice that when I let the mail pile up I waste a lot of time reading junk mail—catalogs of things I'm not going to buy, things I don't need. All of this is odd because now that I am in school time is at a premium and I need to become more efficient."

An example of incident-specific procrastination: Randolph, a 24-year-old counselor trainee, discusses a deferred reconciliation with a cousin who attended the same college and with whom he had a falling out in his senior year. He says, "In high school she had been the popular one. I was a late-bloomer, I suppose. I came to be very well liked in college, while her social life was average. In senior year I found out that she had been telling lies about me behind my back. My success, I guess, galled her. I told her off and our relationship ended. It made family get-togethers difficult. We'd avoid each other, even though our families are close. So I found reasons to avoid family events. But I think we've both matured quite a bit over the last few years. I want to re-establish a relationship with her, but I keep finding reasons to put it off."

1. Describe an instance of chronic or project-specific procrastination in your own life.

_____

_____

_____

_____

_____

_____

2. What are the inhibitors? What keeps you from acting? What are the incentives for *not* acting?

_____

_____

_____

_____

_____

_____

3. What available incentives would help you move to action? What can you do to mobilize them?

_____

_____

_____

_____

_____

_____

4. Share your findings with a learning partner. What do you need to do to move beyond either your habit of procrastinating or the specific instance of procrastination you described?

EXERCISE 92: EXPLORING THE SELF-STARTER IN YOURSELF

While all of us have a tendency to put things off, we also have the possibility of become self-starters. Self-starters move to problem-managing and opportunity-developing action without being influenced,

167

asked, or forced to do so. Skilled helpers are, ideally, self-starters. They have a sense of "agency" that enables them to help clients explore self-starting possibilities in themselves.

Example. Dahlia, whose chronic procrastination is described above, came up with the following list:

- "I never put off the things I like to do. For instance, I like the volunteer work I do at the hospital. No one has to put pressure on me to show up."
- "I make new friends easily. I don't wait for people to approach me. I take the initiative."
- "I plan family get-togethers and make them happen. Family solidarity is an important value for me. I do whatever I can to develop it."
- "I'm in no way a hypochondriac, but I do watch my health carefully. I get yearly check-ups, I see the dentist regularly, and if I experience symptoms that could mean something, I get myself checked out."

1. Describe four ways in which you are a self-starter.

_____

_____

_____

_____

_____

_____

_____

_____

2. Describe two ways in which you would like to become a self-starter.

_____

_____

_____

_____

_____

3. Share what you have learned about yourself as a self-starter with a learning partner. If you have trouble finding ways in which you are a self-starter or ways to become one, explore the implications of this with the training group.

EXERCISE 93: LEARNING FROM FAILURES

As suggested in the text, inertia and entropy dog all of us in our attempts to manage problem situations and develop unexploited opportunities. There is probably no human being who has not failed to carry through on some self-change project. This exercise assumes that we can learn from our failures.

Example: Miguel kept saying that he wanted to leave his father's business and strike out on his own, especially since he and his father had heated arguments over how the business should be run. He earned an MBA in night school and talked about becoming a consultant to small family-run businesses. A medium-sized consulting firm offered him a job. He accepted on the condition that he could finish up some work in the family business. But he always found "one more" project in the family business that needed his attention. All of this came out as part of his story, even though his main concern was the fact that his woman friend of five years had given him an ultimatum: marriage or forget about the relationship.

Finally, with the help of a counselor, Miguel makes two decisions: to take the job with the consulting firm and to agree to break off the relationship with his woman friend because he is still not seriously entertaining marriage as an immediate possibility. However, in the ensuing year Miguel never gets around to taking the new job. He keeps finding tasks to do in his father's company and keeps up his running battle with his father. Obviously both of them, father and son, were getting something out of this in some twisted way. As to his relationship with his woman friend, the two of them broke it off four different times during that year until she finally left him and got involved with another man.

Some of Miguel's learnings. Since Miguel and his counselor were not getting anywhere, they decided to break off their relationship for a while. However, when his woman friend definitively broke off their relationship, Miguel was in such pain that he asked to see the counselor again. The first thing the counselor did was to ask Miguel what he had learned from all that had happened, on the assumption that these learnings could form the basis of further efforts. Here are some of his learnings:

- I hate making decisions that tie me down.
- I pass myself off as an adventuresome, action-oriented person, but at root I prefer the status quo.
- I am very ambiguous about facing the developmental tasks of an adult my age. I have liked living the life of a 17-year-old at age 30.
- I enjoy the stimulation of the counseling sessions. I enjoy reviewing my life with another person and developing insights, but this process involves no real commitment to action on my part.
- Not taking charge of my life and acting on goals has led to the pain I am now experiencing. In putting off the little painful actions that would have served the process of gradual growth, I have ended up in big pain. And I picture myself repeating the same pattern in the future.

Notice that Miguel's learnings are about himself, his problem situations, and his way of participating in the helping process. Now turn to your own life.

1.   Recall some significant self-change project over the last few years that you abandoned in one way or another.
2.   Picture as clearly as possible the forces at work that led to the project's ending in failure, if not with a bang, then with a whimper.
3.   In reviewing your failed efforts, jot down what you learned about yourself and the process of change.
4.   Share your learnings with a learning partner. Help each other: (a) discover further lessons in the review of the failed project, and (b) discover what could have been done to keep the project going.

The self-change project that failed.

_____

_____

_____

_____

Principal reasons for failure.

_____

_____

_____

_____

_____

What you learned about yourself as an agent of change in your own life.

_____

_____

_____

_____

_____

What could have been done to keep the project going?

_____

_____

_____

_____

_____

EXERCISE 94: IDENTIFYING AND COPING WITH OBSTACLES TO ACTION

As suggested in an earlier exercise, "forewarned is forearmed" in the implementation of any plan. Identifying possible obstacles to a constructive change project is wisdom, not weakness.

1.   Picture yourself trying to implement some action strategy or plan in order to accomplish a problem-managing goal. As in the example below, jot down what you actually see happening.
2.   As you tell the story, describe the pitfalls or snags you see yourself encountering along the way. Some pitfalls involve inertia, that is, not starting some step of your plan; others involve entropy, that is, allowing the plan to run into a brick wall or fall apart over time.
3.   Design some strategy to handle any significant snag or pitfall you identify.

Example: Chester has a supervisor at work that, he feels, does not like him. He says that she gives him the worst jobs, asks him to put in overtime when he would rather go home, and talks to him in demeaning ways. In the problem exploration phase of counseling, he discovered that he probably reinforces her behavior by buckling under, by giving signs that he feels hurt but helpless, and by failing to challenge her in any direct way. He feels so miserable at work that he wants to do something about it. One option is to move to a different department, but to do so he must have the recommendation of his immediate supervisor. Another possibility is to quit and get a job elsewhere, but, since he likes the company, that would be a drastic option. A third possibility is to deal with his supervisor more directly. He sets goals related to this third option.

One major step in working out this overall problem situation is to seek out an interview with his supervisor and tell her, in a strong but non-punitive way, his side of the story and how he feels about it. Whatever the outcome, his version of the story would be on record. The counselor asks him to imagine himself doing all of this. What snags does he run into? Here are some of them:

• "I see myself about to ask her for an appointment. I see myself hesitating to do so because she might answer me in a sarcastic way. Also, others are usually around and she might embarrass me and they will want to know what's going on, why I want to see her, and all that. I tell myself that I had better wait for a better time to ask."

• "I see myself sitting in her office. Instead of being firm and straightforward, I'm tongue-tied and apologetic. I forget some of the key points I want to make. I let her brush off some of my complaints and in general let her control the interaction."

How can he prepare himself to handle the obstacles or snags he sees in his first statement? Then what could he do about making sure he gets an appointment?

_____

_____

_____

_____

_____

_____

How can he prepare himself to handle the pitfalls mentioned in his second statement? What could he do to deliver his messages in a decent but forceful way?

_____

_____

_____

_____

_____

_____

Personal Situation # 1.

Consider some plan or part of a plan you want to implement. In your mind's eye see yourself moving through the steps of the plan. What obstacles or snags do you encounter? Jot them down.

_____

_____

_____

_____

_____

_____

Indicate how you might prepare yourself to handle a significant obstacle or pitfall and what you might do in the situation itself to handle it.

_____

_____

_____

_____

_____

_____

Personal Situation # 2.

Consider some plan or part of a plan you want to implement. In your mind's eye see yourself moving through the steps of the plan. What obstacles or snags do you encounter? Jot them down.

_____

_____

_____

_____

_____

_____

Indicate how you might prepare yourself to handle a significant obstacle or pitfall and what you might do in the situation itself to handle it.

_____

_____

_____

_____

_____

_____

EXERCISE 95: IDENTIFYING FACILITATING FORCES

In this exercise you are asked to identify forces "in the field," that is, out there in clients day-to-day lives, that might help them implement strategies and plans, together with forces that might hinder them. The former are called "facilitating forces" and the latter "restraining forces."

Example: Ira, as we have seen earlier, wants to stop smoking. He has also expanded his goal from merely "stopping" to "staying stopped." He has formulated a step-by-step plan for doing so. Before taking the first step, he uses force-field analysis to identify facilitating and restraining forces in his everyday life.

Some of the facilitating forces identified by Ira:

- my own pride.
- the satisfaction of knowing I'm keeping a promise I've made to myself.
- the excitement of a new program, the very "newness" of it.
- the support and encouragement of my wife and my children.
- the support of two close friends who are also quitting.
- the good feeling of having that "gunk" out of my system.
- the money saved and put aside for more reasonable pleasures.
- the ability to jog without feeling I'm going to die.
- becoming more aware of the spiritual dimensions of life.

Now carry out the following tasks:

1. Review a goal or subgoal and the plan you have formulated to accomplish it.
2. Picture yourself "in the field" actually trying to implement the steps of the plan.
3. Identify the principal forces that are helping you reach your goal or subgoal.

Spell out a goal or subgoal you want to accomplish and the key steps you are taking to implement it.

_____

_____

_____

_____

Picture yourself in the process of implementing the plan formulated to achieve the goal. List the facilitating forces that could help you to carry out the plan.

_____

_____

_____

_____

_____

Which facilitating forces are most useful? What can you do to strengthen and use them? For instance, Ira focuses on three related facilitating forces: his pride, the satisfaction keeping commitments he makes, and feeling better by getting the "gunk" out of is system.

- Since he is a religious man, every morning he prayerfully recommits himself to his promise to quit.
- Every evening he congratulates himself for keeping his promises despite "flirtations with temptation."
- He periodically prides himself on becoming less dependent on chewing the gum that satisfies his craving for nicotine.
- He slowly and prudently increases his jogging time and "glories in the fresh air" in his lungs.

Now do the same for your action program.

_____

_____

_____

_____

_____

_____

4.   Finally, through dialogue share your findings with a learning partner.

EXERCISE 96: DEALING WITH RESTRAINING FORCES

Now turn your attention to whatever restraining forces there might be "in the field." When these are weak, they can be brushed aside. However, when they are powerful, they must be dealt with. We return to the case of Ira.

Some of the restraining forces identified by Ira:

- the craving to smoke that I take with me everywhere.
- seeing other people smoke.
- danger times: when I get nervous, after meals, when I feel depressed and discouraged, when I sit and read the paper, when I have a cup of coffee, at night watching television.
- being offered cigarettes by friends.
- when the novelty of the program wears off (and that could be fairly soon).
- increased appetite for food and the possibility of putting on weight.
- my tendency to rationalize and offer great excuses for my failures.
- the fact that I've tried to stop smoking several times before and have never succeeded.

Now do the following tasks:

1. Review a goal or subgoal you identified in the previous exercise.
2. Once more, picture yourself "in the field" actually trying to implement the steps of the plan.
3. Identify the principal forces that are hindering you from reaching your goal or subgoal.

List the restraining forces that might keep you from carrying out the plan.

_____

_____

_____

_____

_____

_____

Which restraining forces are most powerful? What can you do to eliminate or minimize them? For example, Klaus is an alcoholic who wants to stop drinking. He joins Alcoholics Anonymous. During a meeting he is given the names and telephone numbers of two people whom he is told he may call at any time of the day or night if he feels he needs help. He sees this as a critical facilitating force—just knowing that help is around the corner when he needs it. However, Klaus sees being able to get help anytime as a kind of dependency. He hates dependency since he sees that its roots go deep inside him. This is a major restraining force. How does he handle it?

- First, he talks out the negative feelings he has about being dependent in this way with a counselor. That is, the first step in managing this restraining force is to name it.
- In talking, he soon realizes that calling others is a temporary form of dependency that is instrumental in achieving an important goal, developing a pattern of sobriety. He sees calling as a safety value, not a continuation of the dependency streak within him.

- Next he calls the telephone numbers a couple of times when he is not in trouble just to get the feel of doing so. He likes letting others know that he is in good shape.
- He puts the numbers in his wallet, he memorizes them, and he puts them on a piece of paper and carries them in a medical bracelet that tells people who might find him drunk that he is an alcoholic trying to overcome his problem. That is, he eliminates one excuse for not using the numbers.
- Finally, he calls the telephone numbers a couple of times when the craving for alcohol is high and his spirits are low. That is, he gets practice in using this temporary resource.

Now do the same for your change program.

_____

_____

_____

_____

_____

_____

_____

_____

_____

4.   Finally, share your findings with a learning partner. Use empathy, probing, and challenge to help each other learn from the exercise.

EXERCISE 97: USING SUPPORTIVE AND CHALLENGING RELATIONSHIPS

Key people in the day-to-day lives of clients can play an important part in helping them stay on track as they move toward their goals. If part of a client's problem is that he or she is "out of community," then a parallel part of the helping process should be to help the client develop supportive human resources in his or her everyday life. In this exercise you are asked to look at strategies and plans from the viewpoint of these human resources. People can provide both support and challenge.

Example: Enid, a 40-year-old single woman, is trying to decide what she wants to do about a troubled relationship with a man. She knows that she no longer wants to tolerate the psychological abuse she has been getting from him, but she also fears the vacuum she will create by cutting the relationship off. She is, therefore, trying to develop some possibilities for a better relationship. She also realizes that ending the relationship might be the best option. Because of counseling, she has been much more assertive in the relationship. She now cuts off contact whenever he becomes abusive. That is, she is already engaging in a series of "little actions" that help her better manage her life and discover further possibilities. Over the course of two years, the counselor helps Enid develop human resources for both support and challenge.

- She moves from one-to-one counseling to group counseling with occasional one-to-one sessions. Group members provide a great deal of both support and challenge.

- She begins attending church. In the church she attends a group something like Alcoholics Anonymous.
- Through one of the church groups she meets and develops a friendship with a 50-year-old woman who has "seen a lot of life." She challenges Enid whenever she begins feeling sorry for herself. She also introduces Enid to the world of art.
- Enid does some volunteer work at an AIDS center. The work challenges her and there is a great deal of camaraderie among the volunteers. It also takes her mind off herself.

Since one of Enid's problems is that she is "out of community," these human resources constitute part of the solution. All these contacts give her multiple opportunities to get back into community and both give and get support and challenge. Now move on to the following tasks:

1. Summarize some goal you are pursuing and the action plan you have developed to get you there.
2. Identify the human resources that are already part of that plan.
3. Indicate the ways in which people provide support for you as you implement your plan.
4. Indicate ways in which people challenge you to keep to your plan or even change it when appropriate.
5. What further support and challenge would help you stick to your program?
6. Indicate ways of tapping into or developing the people resources needed to provide that support and challenge.

Summarize your goal and action plan.

_____

_____

_____

_____

_____

_____

Identify the human resources that are already part of that plan.

_____

_____

_____

_____

Indicate the ways in which people provide *support* for you as you implement your plan.

_____

_____

_____

_____

_____

_____

Indicate ways in which people *challenge* you to keep to your plan or even change it when appropriate.

_____

_____

_____

_____

_____

What further support and challenge would help you stick to your program?

_____

_____

_____

_____

_____

Indicate ways of tapping into or developing the people resources needed to provide that support and challenge.

_____

_____

_____

_____

_____